*B*iobased Industrial Products

Priorities for Research and Commercialization

Committee on Biobased Industrial Products

Board on Biology

Commission on Life Sciences

National Research Council

NATIONAL ACADEMY PRESS
Washington, D.C.

NATIONAL ACADEMY PRESS • 2101 Constitution Avenue, NW • Washington, DC 20418

NOTICE: The project that is the subject of this report was approved by the Governing Board of the National Research Council, whose members are drawn from the councils of the National Academy of Sciences, the National Academy of Engineering, and the Institute of Medicine. The members of the committee responsible for the report were chosen for their special competences and with regard for appropriate balance.

This report has been prepared with funds provided by the U.S. Department of Agriculture, under agreement number 92-COOP-2-8321; U.S. Department of Energy under order number DE-A101-93CE 50370; National Renewable Energy Laboratory under agreement number XC-2-11274-01; and National Science Foundation under agreement number BCS-9120391. Any opinions, findings, conclusions, or recommendations expressed in this publication are those of the author(s) and do not necessarily reflect the views of the organizations or agencies that provided support for the project.

Library of Congress Cataloging-in-Publication Data

Biobased industrial products : priorities for research and
commercialization / Committee on Biobased Industrial Products, Board on
Biology, Commission on Life Sciences, National Research Council.
    p. cm.
Includes bibliographical references (p.   ) and index.
    ISBN 0-309-05392-7 (casebound)
    1. Biotechnology—United States—Forecasting. 2.
Biotechnology—Government policy—United States. I. National Research
Council (U.S.). Committee on Biobased Industrial Products.
    TP248.185 .B535 1999
    338.4'76606'0973—dc21
                                                                              99-50917

Additional copies of this report are available from the National Academy Press, 2101 Constitution Avenue, NW, Lockbox 285, Washington, DC 20055; (800) 624-6242 or (202) 334-3313 (in the Washington metropolitan area); Internet, http://www.nap.edu

Copyright 2000 by the National Academy of Sciences. All rights reserved.

Printed in the United States of America

THE NATIONAL ACADEMIES

National Academy of Sciences
National Academy of Engineering
Institute of Medicine
National Research Council

The **National Academy of Sciences** is a private, nonprofit, self-perpetuating society of distinguished scholars engaged in scientific and engineering research, dedicated to the furtherance of science and technology and to their use for the general welfare. Upon the authority of the charter granted to it by the Congress in 1863, the Academy has a mandate that requires it to advise the federal government on scientific and technical matters. Dr. Bruce M. Alberts is president of the National Academy of Sciences.

The **National Academy of Engineering** was established in 1964, under the charter of the National Academy of Sciences, as a parallel organization of outstanding engineers. It is autonomous in its administration and in the selection of its members, sharing with the National Academy of Sciences the responsibility for advising the federal government. The National Academy of Engineering also sponsors engineering programs aimed at meeting national needs, encourages education and research, and recognizes the superior achievements of engineers. Dr. William A. Wulf is president of the National Academy of Engineering.

The **Institute of Medicine** was established in 1970 by the National Academy of Sciences to secure the services of eminent members of appropriate professions in the examination of policy matters pertaining to the health of the public. The Institute acts under the responsibility given to the National Academy of Sciences by its congressional charter to be an adviser to the federal government and, upon its own initiative, to identify issues of medical care, research, and education. Dr. Kenneth I. Shine is president of the Institute of Medicine.

The **National Research Council** was organized by the National Academy of Sciences in 1916 to associate the broad community of science and technology with the Academy's purposes of furthering knowledge and advising the federal government. Functioning in accordance with general policies determined by the Academy, the Council has become the principal operating agency of both the National Academy of Sciences and the National Academy of Engineering in providing services to the government, the public, and the scientific and engineering communities. The Council is administered jointly by both Academies and the Institute of Medicine. Dr. Bruce M. Alberts and Dr. William A. Wulf are chairman and vice chairman, respectively, of the National Research Council.

COMMITTEE ON BIOBASED INDUSTRIAL PRODUCTS

CHARLES J. ARNTZEN, *Co-chair*, Boyce Thompson Institute for Plant Research, Inc., Ithaca, New York
BRUCE E. DALE, *Co-chair*, Department of Chemical Engineering, Michigan State University, East Lansing
ROGER N. BEACHY, The Scripps Research Institute, La Jolla, California
JAMES N. BEMILLER, Whistler Center for Carbohydrate Research, Purdue University, West Lafayette, Indiana
RICHARD R. BURGESS, McArdle Laboratory for Cancer Research, University of Wisconsin, Madison
PAUL GALLAGHER, Department of Economics, Iowa State University, Ames
RALPH W. F. HARDY, National Agricultural Biotechnology Council, Ithaca, New York
DONALD L. JOHNSON, Grain Processing Corporation, Muscatine, Iowa
T. KENT KIRK, Forest Products Laboratory, U.S. Department of Agriculture, Madison, Wisconsin
GANESH M. KISHORE, Monsanto Agricultural Group, Chesterfield, Missouri
ALEXANDER M. KLIBANOV, Department of Chemistry, Massachusetts Institute of Technology, Cambridge
JOHN PIERCE, DuPont Agricultural Enterprise, Newark, Delaware
JACQUELINE V. SHANKS, Department of Chemical Engineering, Rice University, Houston, Texas
DANIEL I. C. WANG, Biotechnology Process Engineering Center, Massachusetts Institute of Technology, Cambridge
JANET WESTPHELING, Genetics Department, University of Georgia, Athens
J. GREGORY ZEIKUS, MBI International, Lansing, Michigan

Consultant

Elizabeth Chornesky

Staff

Mary Jane Letaw, *Program Officer*
Joseph Zelibor, *Project Director to January 31, 1996*
Eric Fischer, *Study Director to January 5, 1997*
Paul Gilman, *Study Director to September 30, 1998*

BOARD ON BIOLOGY

PAUL BERG, *Chair*, Stanford University School of Medicine, Stanford, Calif.
JOANNA BURGER, Rutgers University, Piscataway, N.J.
MICHAEL T. CLEGG, University of California, Riverside
DAVID EISENBERG, University of California, Los Angeles
DAVID J. GALAS, Keck Graduate Institute of Applied Life Science, Claremont, Calif.
DAVID V. GOEDDEL, Tularik, Inc., San Francisco
ARTURO GOMEZ-POMPA, University of California, Riverside
CORY S. GOODMAN, University of California, Berkeley
CYNTHIA K. KENYON, University of California, San Francisco
BRUCE R. LEVIN, Emory University, Atlanta, Ga.
ELLIOT M. MEYEROWITZ, California Institute of Technology, Pasadena
ROBERT T. PAINE, University of Washington, Seattle
RONALD R. SEDEROFF, North Carolina State University, Raleigh
ROBERT R. SOKAL, State University of New York, Stony Brook
SHIRLEY M. TILGHMAN, Princeton University, Princeton, N.J.
RAYMOND L. WHITE, University of Utah, Salt Lake City

Staff

Ralph Dell, Acting Director

COMMISSION ON LIFE SCIENCES

MICHAEL T. CLEGG, *Chair*, University of California, Riverside
PAUL BERG, *Vice Chair*, Stanford University School of Medicine, Stanford, Calif.
FREDERICK R. ANDERSON, Cadwalader, Wickersham & Taft, Washington, D.C.
JOHN C. BAILAR III, University of Chicago, Chicago, Il.
JOANNA BURGER, Rutgers University, Piscataway, N.J.
JAMES E. CLEAVER, University of California, San Francisco
DAVID S. EISENBERG, UCLA-DOE Laboratory of Structural Biology and Molecular Medicine, University of California, Los Angeles
JOHN L. EMMERSON, Eli Lilly and Co. (ret.), Indianapolis, In.
NEAL L. FIRST, University of Wisconsin, Madison
DAVID J. GALAS, Keck Graduate Institute of Applied Life Science, Claremont, Calif.
DAVID V. GOEDDEL, Tularik, Inc., South San Francisco, Calif.
ARTURO GOMEZ-POMPA, University of California, Riverside
COREY S. GOODMAN, University of California, Berkeley
JON W. GORDON, Mount Sinai School of Medicine, New York, N.Y.
DAVID G. HOEL, Medical University of South Carolina, Charleston
BARBARA S. HULKA, University of North Carolina at Chapel Hill
CYNTHIA J. KENYON, University of California, San Francisco
BRUCE R. LEVIN, Emory University, Atlanta, Ga.
DAVID M. LIVINGSTON, Dana-Farber Cancer Institute, Boston, Mass.
DONALD R. MATTISON, March of Dimes, White Plains, N.Y.
ELLIOT M. MEYEROWITZ, California Institute of Technology, Pasadena
ROBERT T. PAINE, University of Washington, Seattle
RONALD R. SEDEROFF, North Carolina State University, Raleigh
ROBERT R. SOKAL, State University of New York, Stony Brook
CHARLES F. STEVENS, M.D., The Salk Institute for Biological Studies, La Jolla, Calif.
SHIRLEY M. TILGHMAN, Lewis Thomas Laboratory, Princeton University, Princeton, N.J.
RAYMOND L. WHITE, University of Utah, School of Medicine, Salt Lake City

Staff

Warren Muir, Executive Director

Acknowledgments

This report was reviewed in draft form by individuals chosen for their diverse perspectives and technical expertise in accordance with procedures approved by the National Research Council's Report Review Committee. The purpose of this independent review is to provide candid and critical comments that will assist the institution in making the published report as sound as possible and to ensure that the report meets institutional standards for objectivity, evidence, and responsiveness to the study charge. The review comments and draft manuscript remain confidential to protect the integrity of the deliberative process.

We wish to thank the following individuals for their participation in the review of this report: Margriet Caswell, United States Department of Agriculture Economic Research Service, Washington, D.C.; John S. Chipman, University of Minnesota; Robert E. Connick, retired, University of California, Berkeley; Ronald J. Dinus, retired, University of British Columbia; Raphael Katzen, Consulting Engineer, Bonita Springs, Florida; Scott E. Nichols, Pioneer Hi-Bred International, Inc., Johnston, Iowa; Christopher R. Somerville, Carnegie Institution of Washington, Stanford, California; George T. Tsao, Purdue University; and Charles R. Wilke, retired, University of California, Berkeley.

While the individuals listed above provided constructive comments and suggestions, it must be emphasized that responsibility for the final content of this report rests entirely with the authoring committee and the institution.

Contents

EXECUTIVE SUMMARY 1
   Raw Material Resource Base, 3
   Opportunities: Range of Biobased Products, 5
   Processing Technologies, 8
   A Vision for the Future, 10
   Recommendations, 11

1 OVERVIEW 15
   Potential Benefits of Biobased Industrial Products, 18
      Federal Agricultural Improvement and Reform Act, 19
      International Markets, 19
      Environmental Quality, 19
      Rural Employment, 23
      Diversification of Petroleum Feedstocks, 23
   Setting a Course for the Future, 24
   Report Coverage, 25

2 RAW MATERIAL RESOURCE BASE 26
   Silviculture Crops, 26
   Agricultural Crops, 27
   Enhancing the Supply of Biomass, 29
      Waste Materials, 29
      Conservation Reserve Program, 31

Filling the Raw Material Needs of a Biobased Industry, 32
    Current Resources, 32
    Improving Plant Raw Materials, 39
    Introduction of New Crops, 52
Summary, 53

3   RANGE OF BIOBASED PRODUCTS   55
Commodity Chemicals and Fuels, 57
    Ethanol, 57
    Biodiesel, 58
Intermediate Chemicals, 60
    Ethylene, 60
    Acetic Acid, 62
    Fatty Acids, 62
Specialty Chemicals, 62
    Enzymes, 63
Biobased Materials, 65
    Bioplastics, 66
    Soy-based Inks, 67
    Forest Products, 67
    Cotton and Other Natural Fibers, 68
Targeting Markets, 70
Capital Investments, 71
A Case Study of Lignocellulose-Ethanol Processing, 72

4   PROCESSING TECHNOLOGIES   74
The Biorefinery Concept, 75
    Existing U.S. Prototypes, 75
    Comparison of Biorefineries to Petroleum Refineries, 79
    Lessons from Petroleum Refinery Experience, 80
Processes for Converting Raw Materials to Biobased Products, 81
    Lignocellulose Fractionation Pretreatment: A Key Step, 81
    Thermal, Chemical, and Mechanical Processes, 81
    Biological Processes, 88
Needed Developments in Processing Technology, 95
    Upstream Processes, 95
    Bioprocesses, 96
    Microbiological Systems, 97
    Enzymes, 98
    Downstream Processes, 100
Summary, 101

5 MAKING THE TRANSITION TO BIOBASED PRODUCTS 103
   A Vision for the Future, 104
   Investments to Achieve the Vision, 109
      Niche Products, 110
      Commodity Products, 111
      Public Investments in Research and Development, 111
      Federal-State Cooperation, 113
      Incentives, 113
   Providing a Supportive Infrastructure, 115
      Education of the Public, 115
      Technical Training, 115
      Information and Databases, 116
      Research Priorities, 117
      Biological Research, 117
      Processing Advances, 118
      Economic Feasibility, 123
      Environmental Research, 124
   Conclusion, 124

REFERENCES 126

APPENDIX A: CASE STUDY OF LIGNOCELLULOSE-ETHANOL
PROCESSING 137
   Feedstock Supply and Demand, 137
   Transportation Costs, 140
   Processing Costs, 141
   Fuel Efficiency, 143

APPENDIX B: BIOGRAPHICAL SKETCHES OF COMMITTEE
MEMBERS 144

Tables, Figures, and Boxes

TABLES

2-1	Estimated Availability of Waste Biomass in the United States, 30
2-2	Crops with Potential Uses for Industrial Products, 50
3-1	Increase in Worldwide Sales of Biotechnology Products (1983 and 1994), 56
3-2	Hypothetical Production Cost Comparisons for Ethylene, 61
3-3	Estimated Capital Requirements for Target Biobased Organic Chemicals Produced from Glucose, 72
4-1	Industrial and Food Uses of Corn, 1996 to 1997 Marketing Year, 78
4-2	Comparison of Biorefineries to Fossil-Based Refineries, 80
5-1	Targets for a National Biobased Industry, 105
5-2	Steps to Achieve Targets of a National Biobased Industry: Biobased Liquid Fuels—Production Milestones, 106
5-3	Steps to Achieve Targets of a National Biobased Industry: Biobased Organic Chemicals—Production Milestones, 107
5-4	Steps to Achieve Targets of a National Biobased Industry: Biobased Materials—Production Milestones, 108
A-1	Costs of Corn Stover Harvest in the United States, 1993, 139
A-2	Production Cost Estimate for Plant Processing Corn Stover to Ethanol, 142

FIGURES

1-1	Biobased Products Manufactured Today, 16	
4-1	Corn Processing and Fermentation Chemicals, 76	
4-2	Soybean Processing, 78	
A-1	Corn Stover Supply and Demand Curve, 138	

BOXES

1-1	Converting Biomass to Ethanol, 21	
2-1	Nature's Nylons, 36	
2-2	Evaluating Alternative Crop Sources of Petroselenic Acid, 38	
2-3	Genetic Engineering Methods, 40	
2-4	Genetic Engineering to Increase Starch Biosynthesis, 48	
3-1	Plastics from Plants and Microbes, 66	
3-2	Biopolymers, 69	
4-1	Softening Wood the Natural Way, 89	
4-2	The Changing U.S. Role in Worldwide Amino Acid Production, 91	
4-3	Making Alternative Sweeteners from Corn, 93	

Executive Summary

Biological sciences are likely to make the same impact on the formation of new industries in the next century as the physical and chemical sciences have had on industrial development throughout the century now coming to a close. The biological sciences, when combined with recent and future advances in process engineering, can become the foundation for producing a wide variety of industrial products from renewable plant resources. These "biobased industrial products" will include liquid fuels, chemicals, lubricants, plastics, and building materials. For example, genetically engineered crops currently under development include rapeseed that produces industrial oils, corn that produces specialty chemicals, and transgenic plants that produce polyesters. Except perhaps for large-scale production of bioenergy crops, the land and other agricultural resources of the United States are sufficient to satisfy current domestic and export demands for food, feed, and fiber and still produce the raw materials for most biobased industrial products.

During this century petroleum-based industrial products gradually replaced similar products once made from biological materials. Now, biobased industrial products are beginning to compete with petroleum-derived products that once displaced them. This progress has been made possible by the wealth of knowledge on the scientific basis for conversion of biomass to sugars and other chemicals, particularly the knowledge of biochemical and fermentation fundamentals and related progress in process technology and agricultural economics. New discoveries occurring in microbial, chemical, and genetic engineering research, in particular, could lead to technological advances necessary to reduce the cost of biobased industrial products. Near-term strategies may be dominated by fermentation of sugars through microbial processes for production of commodity chemicals. In the long run, similar processes may be used for large-scale conversion of biomass to liquid fuel. In the future, novel chemicals and materials that cannot be produced from petroleum may be directly extracted from plants. Today only a small fraction of available biomass is used to produce biobased chemicals due to their high conversion costs. The long-term growth of biobased industrial products will depend on the development of cost-competitive technologies and access to diverse markets.

There remains an open question as to the size of petroleum reserves and the future cost of petroleum products. Current oil reserves are substantial, and exploration continues to open new petroleum supplies for the world market (e.g., Caspian Sea). Experts estimate that two-thirds of the world's proven reserves are located in a single geographic region, the Persian Gulf, and that this area will continue to serve as a dominant source for oil exports (USDOE, 1998). Some geologists report that oil reserves could be depleted within 20 years (Kerr, 1998). According to the American Petroleum Institute, there were approximately 43 years of reserves remaining as of 1997 (API, 1997), an increase from the 34 years prevailing before the first Organization of Petroleum Exporting Countries crisis in 1973. While this committee believes there is a need to make a transition to greater use of renewable materials as oil and other fossil fuels are gradually depleted, the committee cannot predict with any accuracy the availability and cost of future supplies of petroleum.

Biobased products have the potential to improve the sustainability of natural resources, environmental quality, and national security while competing economically. Agricultural and forest crops may serve as alternative feedstocks to fossil fuels in order to moderate price and supply disruptions in international petroleum markets and help diversify feedstock sources that support the nation's industrial base. Biobased products may be more environmentally friendly because they are produced by less polluting analogous processes than in the petrochemical industry. Some rural areas should be well positioned to support regional processing facilities dependent on locally grown crops. As a renewable energy source, biomass does not contribute to carbon dioxide in the atmosphere in contrast to fossil fuels. The committee believes that these benefits of biobased products are real. However, these and other benefits listed below have not, in most cases, undergone a rigorous analysis to demonstrate their validity:

- use of currently unexploited productivity in agriculture and forestry;
- reliance on products and industrial processes that are more biodegradable, create less pollution, and generally have fewer harmful environmental impacts;
- development of less expensive and better-performing products;
- development of novel materials not available from petroleum sources;
- exploitation of U.S. capacities in the field of molecular biology to selectively modify raw materials and reduce the costs of raw material production and processing;
- revitalization of rural economies by production and processing of renewable resources in smaller communities;
- reduction of the potential for disruption of the U.S. economy due to dependence on imported fuel;
- countering of oligopoly pricing on world petroleum markets; and
- mitigation of projected global climate change through reduction of buildup of atmospheric carbon dioxide.

The committee believes that these potential benefits could justify public policies that encourage a transition to biobased industrial products. This report identifies promising resources, technologies, processes, and product lines. Ultimately, the decision as to whether to accelerate investment in the research and development of cost-competitive biobased industrial products will be made by policymakers.

RAW MATERIAL RESOURCE BASE

The United States is well prepared to supply industrial production's growing demand for biological raw materials. The country has abundant croplands and forests, favorable climates, accessible capital, and a skilled labor force that uses sophisticated technologies in agriculture and silviculture. The expansion of biobased industries will depend on currently unused land and byproducts of U.S. agriculture and forestry, on expected increases in crop productivity, and on coproduction of biobased products with traditional food, feed, and fiber products. Enough waste biomass is generated each year—approximately 280 million tons—to supply domestic consumption of all industrial chemicals that can readily be made from biomass and also contribute to the nation's liquid transportation fuel needs. Productivity of U.S. farms and forests has been rising to meet domestic and export demands for traditional food, feed, and fiber products as well as biobased raw materials. Approximately 35 million acres of marginal cropland in the Conservation Reserve Program could provide additional land to grow biomass crops. If approximately half of the land set aside for the program could be harvested in a judicious manner (to minimize the risks of soil erosion and loss of wildlife), approximately 46 million tons of additional biomass feedstock would become available. This figure assumes very low yields of biomass (2.5 tons per acre) and could increase fourfold (up to 10 tons per acre) with some crops (e.g., switchgrass). The total biomass is sufficient to easily meet current demands for biobased industrial chemicals and materials.

The amount of land that will actually be used for biobased crops will depend on future demands for the final products, and the inputs used to make those products must be competitively priced. High-value novel chemicals are not expected to require large acreages. While biobased materials such as lumber, cotton, and wool do have substantial markets, these products now compete successfully for land resources. However, the current demand for many biobased chemical products is small. For example, as of the 1996 to 1997 marketing year, industrial uses of starch and manufacturing and fuel ethanol production from corn accounted for approximately 7 percent of the nation's corn grain production (ERS, 1997b).

Coproduction of human food and animal feed products such as protein with biobased fuels, chemicals, and materials is expected to help minimize future conflicts between production of food and biobased products. Corn-based refineries, for example, yield protein for animal feed and oil, starch, fiber, and fuel alcohol products. In the case of pulp and paper mills, pulp, paper, lignin byproducts, and ethanol can be produced while recycling waste paper in a single system. If demand for liquid fuel increases beyond capacity for coproduction of food and liquid fuel, biobased production may compete for land with food production. This report describes some opportunities for coproduction of food, feed, liquid fuels, organic chemicals, and materials.

The committee recognizes that an abundant supply of food at a reasonable price is a national goal. If the oil supply does diminish without available substitutes, oil prices could rise. At that point, policymakers may decide to convert land from food to fuel production. This could create competition for scarce resources and subsequent conversion of U.S. croplands to energy crops could lead to higher food prices. The committee estimates that byproducts of agriculture could provide up to 10 percent of liquid transportation fuel needs. The amount of land devoted to crops for biobased industries will be determined by economics, as tempered by agricultural policies.

The raw materials for biobased industrial production are supplied by plants as plant parts, separated components, and fermentable sugars. For the immediate future the raw material sources most likely to be used for producing industrial materials and chemicals in the United States are starch crops like corn and possibly waste biomass. Over the long term, as the demand for biobased products expands and crop conversion technologies improve, this resource base will grow to include lignocellulosic materials from grasses, trees, shrubs, crop residues, and alternative crops custom engineered for specialized applications.

Many potential biobased products will come from traditional crop plants being put to new uses—for example, grasses and legumes used in paper production. Perhaps more important, however, will be new types of crops or traditional crops that have been genetically engineered. Although a number of barriers can impede the introduction of new crops, the transformation of soybean from a minor crop earlier in this century to a major crop today illustrates the possibilities when crop production and conversion technologies are developed in tandem.

Genetic engineering and plant breeding techniques permit the redesign of crops for easier processing and creation of new types of raw materials. Source plants can be modified or selected for characteristics that enhance their conversion to useful industrial products. Through genetic engineering, plant cellular processes and components can be altered in ways that increase the value or uses of the modified crop. This capability has no parallel in petroleum-based feedstock systems and is a major advantage of biobased industrial products.

OPPORTUNITIES: RANGE OF BIOBASED PRODUCTS

Biobased products fall into three categories: commodity chemicals (including fuels), specialty chemicals, and materials. Some of these products result from the direct physical or chemical processing of biomass—cellulose, starch, oils, protein, lignin, and terpenes. Others are indirectly processed from carbohydrates by biotechnologies such as microbial (e.g., fermentation) and enzymatic processing. Fermentation ethanol and biodiesel are examples of biobased fuels. Ethanol is critical because this oxygenate can serve as a precursor to other organic chemicals required for production of paints, solvents, clothing, synthetic fibers, and plastics. While ethanol currently is the largest-volume and probably cheapest fermentation product, other chemicals such as lactic acid are under development as raw materials for further processing. Some biobased chemicals are becoming price and cost competitive. For example, vegetable-oil-based inks and fatty acids now account for 8 and 40 percent of their respective domestic markets. Biobased chemicals (apart from liquid fuels) probably represent the greatest near-term opportunity for replacement of petrochemicals with renewable resources.

The driving force for production of many biobased chemicals and liquid fuels has been a search for alternatives to fossil fuels in response to the oil crisis of the 1970s, a desire to reduce stocks of agricultural commodities, and more recent attention to the environment. In many cases, biobased products received a premium price or subsidy when they were introduced to the marketplace. For instance, fermentation ethanol gained a 1 percent share of the domestic transportation fuel market (about 1 billion gallons of ethanol) in 1995 due, in part, to government incentives designed to improve air quality in some urban areas. As more cities meet carbon monoxide air quality standards, this ethanol market will decrease. To penetrate larger commercial markets, ethanol and other commodity chemicals will have to become cost and price competitive with petroleum-based products. Increasingly, technological advances in production processes (as outlined in this report) have the potential to drive down the costs of biobased products, allowing them to compete in an open market with petroleum-derived products.

The worldwide market for specialty chemicals—enzymes, biopesticides, thickening agents, and antioxidants—is $3 billion and growing by 10 to 20 percent per year. The market for detergent enzymes alone is about $500 million annually. As sales volume has increased, the cost of detergent enzymes has fallen 75 percent over the past decade. Based on industry experience, a similar pattern can be expected for other biobased products. Many new applications for enzymes are being explored, including animal feeds, wood bleaching, and leather manufacture. In each, enzymes improve the industrial process and make it less polluting. Increasingly, niche markets will be sought for a wide array of plant chemicals (e.g., chiral compounds) not available from petrochemical markets.

Biobased materials represent a significant market with a wide range of products. Lumber, paper, and wood products have traditionally been a large market, with annual sales of approximately $130 billion in the United States. Several other biobased materials have established uses that are likely to grow as technological advances reduce costs. Examples include starch-derived plastics, biopolymers for secondary oil recovery, paper, and fabric coatings.

The cost of large-scale production of biobased products depends on two primary factors: the cost of the raw material and the cost of the conversion process. The industries for producing chemicals and fuels from petroleum are characterized by high raw material costs relative to processing costs, while in the analogous biobased industries processing costs dominate. Therefore, similar percentage improvements in processing costs have much more impact on biobased industries. Also, the cost per ton of biomass raw materials generally is comparable (e.g., corn grain) or much less (e.g., corn stover) than the cost per ton of petroleum. Thus, there is real potential for biobased products to be cost competitive with petroleum-based products if the necessary research and development are done to reduce processing costs.

Furthermore, because starch and sugar already contain oxygen and petroleum does not, there is the potential to derive oxygenated intermediate chemicals—such as ethylene glycol, adipic acid, and isopropanol—more readily from biological raw materials than from fossil sources. Production of such oxygenated chemicals by fermentation has the additional advantage of being inherently flexible. The raw materials can vary depending on which local source of fermentable sugars provides the best economic returns. Therefore, economic evaluations should first consider the potential of biobased replacements for the oxygenated organic chemicals of the 100 million metric tons of industrial chemicals marketed each year in the United States.

Other significant opportunities exist to produce a wide range of industrial products from agricultural and forest resources. Many will require investment in basic research as well as process engineering research to ensure commercial viability. These opportunities begin with the plant sources for raw materials. Modern principles of molecular biology and genetic engineering can be used to create agricultural crops that contain desired chemical polymers or polymer intermediates. Additionally, trees and grasses could be genetically engineered to have a structural composition that facilitates and enhances the effectiveness and efficiency of subsequent conversion to desired products.

Combined advances in functional genomics, genetic engineering, and biochemical pathway analysis, sometimes referred to as metabolic engineering, will make it possible to manipulate efficiently the biosynthetic pathways of microorganisms. By increasing chemical yield and selectivity, such manipulations could make microbial production more economically competitive with existing production methods. The combination of modern genetics and protein engineering will provide biocatalysts for improved synthesis or conversion of known products or for reaction routes to new chemicals.

Accelerating the growth of biobased products will require an awareness of the opportunities and focused investment in research and development. The pathway to many industrial products starts with basic research. Such research generates promising discoveries that must be proven at a sufficiently large scale to reduce the risks of investing in the final commercial application. Barriers do exist in bridging the gap between laboratory discovery and product commercialization. Industry experience suggests that for every million dollars spent in basic discovery-oriented research for a specific product, $10 million must be spent in the proof-of-concept stage and $100 million in the final commercial-scale application.

Public and industrial investment in basic research in the United States has traditionally been strong and should continue. Final commercialization has been and should remain the province of industry. However, there is limited venture capital that is available for early commercialization of biobased products. This committee believes that the nation could benefit from government-industry partnerships that focus resources on the essential intermediate stage of proof of concept (risk reduction). The degree of public investment in biobased industrial products from basic research through proof of concept will be a public policy decision.

Public risk capital is a mechanism that is currently used to support this intermediate proof-of-concept stage. The Alternative Agricultural Research and Commercialization Corporation administered by the U.S. Department of Agriculture is specifically devoted to commercializing industrial uses of renewable raw materials. A basic tenet of these partnerships is that upon successful commercialization the rate of return of a public investment should be commensurate with other risk capital investments. The public sector also has invested in several demonstration facilities that could support future proof-of-concept activities. Examples include the National Renewable Energy Laboratory (U.S. Department of Energy), the National Center for Agricultural Utilization and Research (U.S. Department of Agriculture), and MBI International (Lansing, Michigan). Such facilities handle a wide range of flexible large-scale processing equipment and have ample qualified support personnel. This committee believes that these facilities should be required to obtain a significant fraction of their funds for demonstration and risk reduction activities from the private sector.

PROCESSING TECHNOLOGIES

The U.S. capacity to produce large quantities of plant material from farms and forests is complemented by the nation's technical capability to convert these plant materials into useful products. Various thermal, chemical, mechanical, and biological processes are involved. Expansion of biobased industrial production in the United States will require an overall scale-up of manufacturing capabilities, diversification of processing technologies, and reduction of processing costs. The development of efficient "biorefineries"—integrated processing plants that yield numerous products—could reduce costs and allow biobased products to compete more effectively with petroleum-based products. Prototype biorefineries already exist, including corn-wet mills, soybean processing facilities, and pulp and paper mills.

As in oil refineries, biorefineries would yield a host of products that would tend to increase over time. Many biorefinery products can be produced by petroleum refineries, such as liquid fuels, organic chemicals, and materials. However, biorefineries can also manufacture many other products that oil refineries cannot, including foods, feeds, and biochemicals. These additional capabilities give biorefineries a potential competitive edge and enhanced financial stability.

The processing technologies of refineries tend to improve incrementally over time, eventually causing raw material costs to become the dominant cost factor. In this regard, biorefineries have another potential advantage over petroleum refineries because plant-derived raw materials are abundant domestic resources. The availability and prices of biological raw materials may thus be more stable and predictable than those of petroleum.

An extensive case study in this report examines the potential of converting corn stover (stalks, leaves, cobs, and husks—also known as corn residue) to ethanol. The case study incorporates a model to calculate costs for ethanol processed from corn stover. Today, production of cornstarch-based ethanol costs approximately $1.05 per gallon. The model indicates that by using corn residue as a feedstock up to 7.5 billion gallons of ethanol could be produced at a cost potentially competitive with gasoline without subsidies. When the ethanol price is adjusted to account for the fact that a gallon of ethanol will provide less mileage in a conventional gasoline-type engine than will the fuel for which the engine is designed, the price of ethanol equivalent to a gallon of gasoline is $0.58 per gallon. The U.S. refinery price for motor gasoline in July 1998 was $0.54 per gallon (EIA, 1998). The model assumes that some not yet completely developed technologies are available and that use of corn residue makes possible especially low-cost raw materials. As a result, projected costs for ethanol processing could drop significantly from current costs because these residues are coproduced with corn grain. It should be noted that the price of oil could change significantly from today's prices, thus changing the price comparisons between ethanol and gasoline. The opportunities to produce ethanol more efficiently are large. While corn has been the dominant raw material source, other more productive lignocellulosic materials such as switchgrass are being considered as alternative feedstocks.

In many cases the biorefinery that produces ethanol and other commodity chemicals from lignocellulosic biomass requires three major new technologies: (1) an effective and economical pretreatment to unlock the potentially fermentable sugars in lignocellulosic biomass or alternative processes that enable more biomass carbon to be converted to ethanol or other desired products; (2) inexpensive enzymes (called "cellulases") to convert the sugar polymers in lignocellulose to fermentable sugars; and (3) microbes that can rapidly and completely convert the variety of 5- and 6-carbon sugars in lignocellulose to ethanol and other oxygenated chemicals.

Several lignocellulose pretreatment processes have recently been developed that promise to be technically effective and affordable. Such pretreatments should make it possible to convert a vast array of lignocellulose resources into useful products. Other biobased processes under development may not require all of these pretreatment processes. Considerable progress has also been made in developing genetically engineered microorganisms, which utilize both 5- and 6-carbon sugars. Less progress apparently has been made in producing inexpensive cellulases.

Processing technologies that use microbes and enzymes have great promise for the expansion of biobased industries. Unlike thermal and chemical processes, such bioprocesses occur under mild reaction conditions, usually result in stereospecific conversions, and produce only a few relatively nontoxic byproducts. One drawback is that bioprocesses typically yield dilute aqueous product streams, requiring subsequent processing for separation and purification. Bioprocessing research should therefore focus on increasing processing rates, product yields, and product concentrations with the overall goal of significant cost reduction. Some advanced bioprocessing concepts have already been developed, such as immobilized cell technology and simultaneous saccharification and fermentation.

Experience with commercial amino acid production demonstrates the advantages of combining inexpensive raw materials with advanced bioprocessing methods. International amino acid markets were completely dominated by Japanese firms in the early 1980s. However, starting in the 1990s, U.S. companies using inexpensive corn-based sugars and immobilized cell technology began to penetrate these markets and today are major players in the industry.

In general, research on the underlying production processes should focus on the science and engineering necessary to reduce the most significant cost barriers to commercialization. Economic and market studies could help clearly identify these barriers, determine the costs of alternative plant feedstocks, and understand the effects of fluctuating industrial demand and agricultural production on the risks and returns for bioprocessing investments. There are also storage and transportation problems unique to biobased products. Most biomass crop production takes place during a portion of the year, but biomass raw materials should be available on a continuous basis for industrial processing. Thus, there is a need to do research in these areas.

A VISION FOR THE FUTURE

The committee has described circumstances that it believes will accelerate the introduction of more sustainable approaches to the production of industrial chemicals, liquid fuels, and materials. In this vision a much larger and competitively priced biobased products industry will eventually replace much of the petrochemical industry. The committee proposes the following intermediate- and long-term targets for the biobased products industry:

- by the year 2020, provide at least 25 percent of 1994 levels of organic carbon-based industrial feedstock chemicals and 10 percent of liquid fuels from a biobased products industry;
- eventually satisfy over 90 percent of U.S. organic chemical consumption and up to 50 percent of liquid fuel needs with biobased products; and
- form the basis for U.S. leadership of the global transition to biobased products and potential environmental benefits.

These targets are based on estimates of available feedstocks and assume that technological advances are in place to improve the suitability of raw materials and the economics of the conversion processes. Ultimately, the extent of this will be determined by the rate of investment by the private sector.

The end of the next century may well see many petroleum-derived products replaced with less expensive, better-performing biobased products made from renewable materials grown in America's forests and fields. The committee believes that movement to a biobased production system is a sensible approach for achieving economic and environmental sustainability. While it is outside this committee's charge to determine the degree of involvement by the public sector in these activities, there may be a compelling national interest to make this transition to biobased industrial products. For example, policymakers may want to accelerate the use of renewable biomass to mitigate adverse impacts on the U.S. economy from a disruption in world oil supplies or reduce adverse impacts on the environment such as those created by possible global warming.

RECOMMENDATIONS

Federal support of research on biobased industrial products can be an effective means of improving the competitiveness of biobased feedstocks and processing technologies, as well as diversifying the nation's industrial base of raw materials and providing additional markets for farmers. Policymakers should encourage research and development that would fill important technical gaps in raw material production, storage, marketing, and processing techniques. Volatility in petroleum prices is a barrier to the development of these biobased products by the private sector.

Policymakers should realize that decades of research investment may be necessary to develop enabling technologies, and considerable lead time will be necessary to implement such research programs and to allow for the adoption of new technologies by industry.

Research will be a prominent tool in making biobased feedstocks more competitive. The public-sector research and development agenda should emphasize major technical and economic roadblocks that impede the progress of biobased industrial products. Research priorities should emphasize the development of biobased products that can compete in performance and cost with fossil-based ones. Expansion of biobased industries will require research on the biological and engineering principles that underlie biobased technologies as well as the practical implementation of these technologies through development and commercialization.

The discoveries occurring today in plant and microbial genomics are expected to lead to significant advances in fundamental biological research for many years in the future. The complete genomic sequence is available for some microbial organisms such as *Saccharomyces cerevisiae* (common yeast) and *Escherichia coli* (gram-negative bacteria). Scientific investigations are under way to decipher the entire genetic code of eukaryote organisms such as *Arabidopsis thaliana* (flowering plant of the mustard family) and *Drosophila* (fly). The genetic information collected on these organisms will provide researchers with insights on the genes that control plant traits and microbial cellular processes. In the future this genomic knowledge will help scientists find new ways to alter microbes and plants that increase the value of biobased raw materials and improve the efficiency of the conversion processes.

Specific recommended research priorities for biology include:

- the genetics of plants and bacteria that will lead to an understanding of cellular processes and plant traits;
- the physiology and biochemistry of plants and microorganisms directed toward modification of plant metabolism and improved bioconversion processes;
- protein engineering methods to allow the design of new biocatalysts and novel materials for the biobased industry; and
- maximization of biomass productivity.

Recommended research priorities for engineering include:

- equipment and methods to harvest, store, and fractionate biomass for subsequent conversion processes;
- methods to increase the efficiency and significantly reduce the costs of conversion of biomass to liquid fuel and organic chemi-

cals, including pretreatment of lignocellulosics, as well as other alternative processes so as to make biobased feedstocks economically competitive;
- principles and processing equipment to handle solid feedstocks;
- fermentation technologies to improve the rate of fermentation, yield, and concentration of biobased products; and
- downstream technologies to separate and purify products in dilute aqueous streams.

Most biologically based technologies and products have the potential to be more benign to the environment than petroleum-based sources. Growing plant matter such as perennial grasses for conversion to industrial products actually has the potential to improve soil quality. The use of biobased products in place of fossil materials does not add to atmospheric carbon dioxide, whereas use of the latter does. With rapidly increasing energy demands in developing nations, the substitution of biomass-derived fuels for fossil fuels could help reduce loading of atmospheric carbon dioxide and its possible impacts on global climate. Many biobased industrial products may prove to be more biodegradable and less polluting and many generate less hazardous wastes than fossil fuels. However, in many cases these benefits have been demonstrated for only a single step of the manufacturing process or for a single emission. Thus, more research in this area is needed. Evaluations of the potential environmental benefits of biobased industrial products should include life-cycle assessments that examine all phases from production and processing of raw materials to waste disposal.

The committee envisions a government-industry partnership in which the public sector facilitates and supports research and in key cases where industry will not risk sole responsibility the government (federal, state, and local) may be a joint supporter of proof of concept. These partnerships should emphasize enabling technologies that are essential to the development of new products and processes across several industries and in cases where there is no other funding source (NRC, 1995). Equally important will be educational support and training to prepare a technical work force able to develop new biobased processes and products.

Biobased industrial development across the United States often will be region or state specific because of differences in agriculture or forestry resources. Consequently, a diversity of approaches to the research, development, and early commercialization of biobased industries is encouraged. Flexible mechanisms to encourage cooperation between federal and state governments, such as matching funds, could help achieve this goal.

Government agencies may decide to implement incentive programs as a mechanism to catalyze biobased industries because the adoption of biobased products will require changes to established industry and consumer practices. For example, a seal of authenticity could create consumer awareness of biobased products and their accompanying environmental benefits. National environmental achievement awards could recognize and reward industry achievements in this area. Other possibilities include tax, investment, and regulatory policies that encourage biobased industries through entrepreneurship and small business formation or that incorporate biobased products into national policies to meet environmental goals. Incentive programs can have widespread implications for the economy and these effects should be carefully considered by government agencies in developing public policies for biobased industrial products. Because the costs of financing some of these incentives are not well known, government agencies will need to obtain comprehensive cost-benefit data for their decisionmaking. Incentive programs should be cost effective with endpoint provisions to evaluate program utility. In the long term, development of biobased products that can compete in an open market without incentives is key to sustaining a strong biobased industry.

Although policy changes would go a long way in encouraging the development of U.S. biobased industries, they will not be sufficient alone. The current technology base for biobased industries is incomplete. Advances in agriculture have stressed crop production technologies without a comparable interest in conversion technologies to produce biobased industrial products. Likewise, education and research resources in the fields of chemistry and process engineering will need to put more emphasis on biobased processing.

This report takes a broad look at current and potential biobased industries. It identifies key opportunities for products derived from renewable resources and the industry and public policy actions that could facilitate the research, development, and commercialization of biobased industrial products. With a vigorous commitment from all parties, the United States will be well positioned to reap the benefits of a strong biobased industry.

1

Overview

Materials that contain carbon play an integral role in the U.S. and world economies. Included here are the primary fuels in commerce, virtually all food and fiber products, and the major share of commodity chemicals, pharmaceuticals, and nondurable manufactured goods. These products are derived from carbon-rich raw materials. The raw materials, in turn, originate through the process of photosynthesis in which plants and some bacteria use solar energy to convert atmospheric carbon dioxide into organic substances, such as sugars, polysaccharides, amino acids, proteins, and fats. Some carbon-rich raw materials come from fossil sources such as petroleum, coal, and natural gas. Fossil sources are the result of photosynthesis in ancient times and comprise a large, but limited, reserve that cannot be renewed. The present-day photosynthesis of plants provides a different living source of carbon. Unlike fossil sources, these biological carbon sources are a potentially renewable asset that is replenished daily by photosynthetic activity.

Renewable agricultural and forestry resources have been used since ancient times as the raw materials for numerous products. For example, Egyptians extracted oil from the castor bean to use as lamp fuel. A shift to fossil sources occurred in the early 1800s, when coal came to dominate U.S. fuel and gas markets and technologies were developed to manufacture chemicals from coal tar. By 1920 chemical producers began using petroleum, and gradually most industries switched from biological raw materials to fossil fuel resources. By the 1970s, organic chemicals derived from petroleum had largely replaced those derived from plant matter,

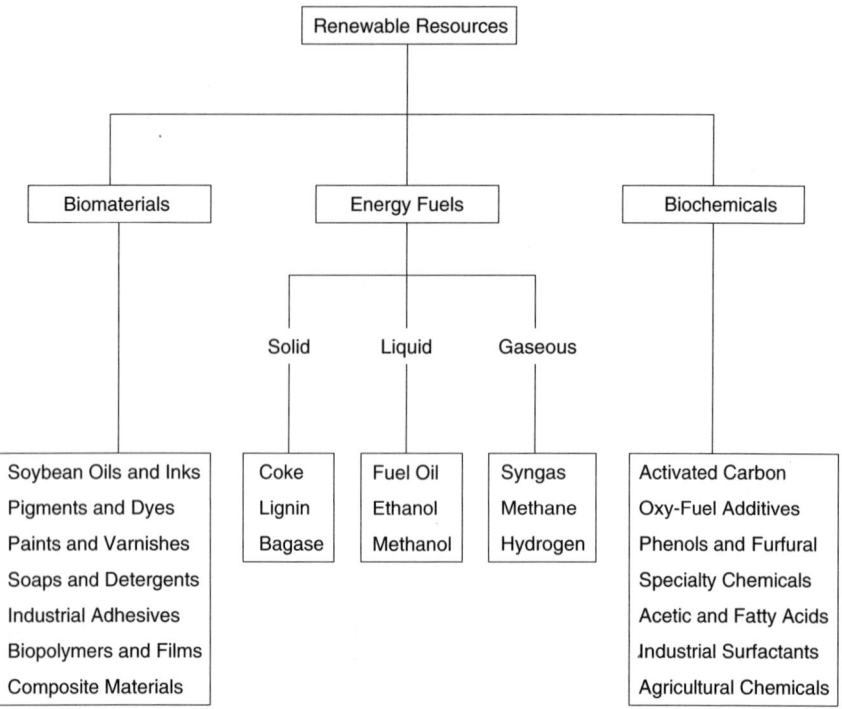

FIGURE 1-1 Biobased products manufactured today. *Source:* Morris and Ahmed (1992). Reprinted with permission.

capturing more than 95 percent of the markets previously held by products made from biological resources, and petroleum accounted for more than 70 percent of our fuels (Morris and Ahmed, 1992). While petroleum does dominate today's industry, there has always been a strong interest in converting underutilized biological materials into useful products (Figure 1–1).

The conversion of agricultural and forest biological raw materials into value-added industrial products continues to be a promising area of research. In the 1970s an embargo organized by the Organization of Petroleum Exporting Countries (OPEC) ignited a period of uncertainty for the United States and generated renewed interest in biobased raw materials. Consequently, U.S. policymakers directed some research funding for development of alternative energy sources that could substitute for fossil fuels. At the same time, public concern for the environment grew and biobased technologies were considered potential replacements for more polluting industrial processes. Today, widespread commercialization of these products has been somewhat limited due to their high cost and lack of viable markets. The remarkable discoveries taking place in the life sciences raise prospects that economically competitive production of more biobased industrial products will be achievable in the future.

In 1995 the National Research Council convened a committee under the Board on Biology to identify priorities for research and commercialization of biobased industrial products derived from agricultural and forestry resources. Committee members were selected for expertise in several key areas, including biomaterials, bioprocessing, economics, enzymology, forest products, lipid and carbohydrate chemistry, microbial and plant genetics, plant biochemistry and pathology, microbiology, and technology transfer. The committee examined the opportunities offered in three areas: (1) recent advances in biotechnology and chemical and material sciences, (2) increases in U.S. agricultural and forest production capacity, and (3) the advantages to the U.S. economy of enhancing industrial growth in rural areas through biobased products. Food and feed products were not considered by the committee, nor were pharmaceuticals.

Most biobased raw materials are produced in agriculture, silviculture, and microbial systems. Silviculture crops are an important source of material for the pulp, paper, construction, and chemical industries. Agricultural crops are chemical feedstocks that can be converted to fuels, chemicals, and biobased materials. Waste biomass should be considered as another major currently unused source of raw materials for U.S. biobased industries. Some biobased industrial products result from direct physical or chemical processing of biomass materials—cellulose, starch, oils, protein, lignin, and terpenes. Others are indirectly produced from carbohydrates by biotechnologies such as microbial and enzymatic processing (Szmant, 1987). Great opportunities now exist to change the raw material focus of our carbon-dependent industries—including energy production and nondurable manufacturing as well as some durable manufacturing.

The relative importance of fossil versus biological carbon sources varies among commercial sectors, as does the potential for expanded reliance on biological carbon sources. Fuels make up about 70 percent of the carbon consumed annually in the United States (1.6 billion to 1.8 billion tons). Biobased fuels, such as ethanol and biodiesel, account for less than 1 percent of total liquid fuel consumption because they are currently more expensive than fossil fuels. Development of low-cost biological carbon sources (e.g., wastes or cellulose biomass) and low-cost high-yield processes will be essential for biobased liquid fuels to become price competitive without subsidization and expand beyond niche markets.

One hundred million metric tons of fine, specialty, intermediate, and commodity chemicals are produced annually in the United States. Only

10 percent of these chemicals are biobased. At present, markets exist for only a few chemicals produced from biological resources such as citric acid, amino acids, sorbitol, and fatty acids. Improved processing technologies and sufficiently low-cost biological carbon feedstocks must be achieved to make production of numerous other chemicals economically competitive.

About 90 percent of the carbon-containing materials (other than chemicals or fuels) in commerce (e.g., lumber and paper, natural polymers and fibers, and composites) are biobased. Lumber and paper products account for well over half of this category; natural polymers or fibers (e.g., cotton), other cellulosics (e.g., rayon, lyocell, and acetate), and certain proteins also are significant. These products are directly extracted from existing crops and trees but in the longer term could be produced from plants or microbes genetically engineered to manufacture specific substances. Research already is under way to biodesign plants to produce biodegradable polyester.

POTENTIAL BENEFITS OF BIOBASED INDUSTRIAL PRODUCTS

Significant benefits could accrue to the United States by switching some production currently dependent on fossil resources to biological sources. This committee identified some potential benefits of biobased industrial products that it believes are real. However, these benefits, which are listed below, have not in most cases undergone a rigorous analysis to demonstrate their validity:

- use of currently unexploited productivity in agriculture and forestry;
- reliance on products that are more biodegradable and processes that create less pollution and generally have fewer harmful environmental impacts;
- development of less expensive and better-performing products;
- development of novel materials not available from petroleum sources;
- exploitation of U.S. capacities in the field of molecular biology to selectively modify raw materials and reduce costs of raw materials production and processing;
- revitalization of rural economies by production and processing of renewable resources in smaller communities;
- reduction of the potential for disruption of the U.S. economy due to dependence on imported fuel;
- countering of oligopoly pricing on world petroleum markets; and

- mitigation of projected global climate change through reduction of buildup of atmospheric carbon dioxide.

Federal Agricultural Improvement and Reform Act

The Federal Agricultural Improvement and Reform Act of 1996 marks a significant change in U.S. agricultural policy (ERS, 1996c). The new law (Public Law 104-127) removes the link between income support payments and farm prices and moves agriculture away from government control toward a market orientation. The legislation authorizes reductions in federal outlays to the farm sector over the years 1996 to 2002. Farmers will have much more flexibility in making planting decisions because of the elimination of annual acreage idling programs and options to plant any crop on contract acres. As a result, producers will rely more heavily on the market as a guide for production decisions (ERS, 1996c).

International Markets

Trends in U.S. policy toward liberalized trade may increase opportunities for exports of biobased products. The 1994 Uruguay Round and the new World Trade Organization reversed long-held policies of protectionism and government control (Roberts, 1998). These agreements are stimulating reform in global trading systems by increasing access to international markets and establishing new rules for freer trade. Trade agreements allow U.S. farmers to better realize competitive gains from their comparative advantage in many agricultural products while reinforcing the advantages of freedom to respond to market signals (USDA, 1997b).

Nations that are technological innovators generally capture the greatest market share, lead in intellectual property and know-how, and create the essential technology platform for further development and innovation. To the extent that biobased fuels can slow global warming, the United States could develop processes for making biomass fuels and market these technologies internationally.

Environmental Quality

Use of fossil carbon sources poses a number of potential hazards to the environment and public health. Chemicals that pollute the air, water, and soil can be released during combustion, processing, or extraction of fossil fuels. The concentration of oil refineries along coasts and rivers creates opportunities for oil spills and their attendant impacts on the environment and wildlife. Use of fossil fuels also releases carbon that was sequestered long ago by photosynthesis and thereby contributes to the worldwide increase in atmospheric carbon dioxide and potentially to global warming. On the other hand, biobased fuels and chemicals are derived from plant materials and can reduce loading of atmospheric carbon dioxide. While this report identifies some biobased products or processes with documented environmental benefits, the environmental benefits (or costs) of most biobased products compared to fossil-based sources are not well known.

On a global scale, there is little doubt that human activities associated with fossil fuels have altered the composition of atmospheric gases (NRC, 1992). Greenhouse gases such as carbon dioxide have increased one-third over preindustrial levels. While considerable debate and research continue on the magnitude and distribution of greenhouse gases and their consequences on humans and the environment, many scientists believe that greenhouse gas emissions will lead to increased global temperatures and associated climate changes (Dixon et al., 1994; USDOE, 1998). Under the United Nations Framework Convention on Climate Change (FCCC), over 150 signatory nations pledged to "adopt policies that limit greenhouse gas emissions and to protect and enhance greenhouse gas sinks and reservoirs." On December 10, 1997, international parties adopted the Kyoto Protocol to the United Nations FCCC to reduce greenhouse gas emission. U.S. administration officials pledged to reduce key greenhouse gases 7 percent below 1990 levels by the period 2008 to 2012. The U.S. Senate has not ratified the agreement (http://www.cop3.de). Biobased fuels could have a significant role in meeting these commitments.

Because biobased fuels, such as alcohol, are derived from renewable (plant) sources (see Box 1-1), they do not add to the carbon dioxide content of the atmosphere, unlike fuels derived from fossil sources (oil, natural gas, coal). When plants are harvested and converted to a biobased fuel, which then is burned, the carbon of the fuel will go into the atmosphere as carbon dioxide. But new plants will now grow and through photosynthesis remove essentially the same amount of carbon dioxide from the atmosphere. This cycle of growth and harvesting can then be repeated indefinitely, with the net production of biobased fuel but no net addition of carbon dioxide to the atmosphere.

Plants can also act as a sink for carbon dioxide. Trees as they grow store increasing amounts of carbon. If they are harvested, however, and converted to fuel, their carbon is once again returned to the atmosphere. Therefore, the effectiveness as a carbon sink of fast-growing and quickly harvested trees is quite limited. With rapidly increasing energy demands in the Third World countries, fossil fuels could make potential global warming eventually very disruptive, unless nonfossil sources can be substituted. The evaluations of biomass energy system effects on atmospheric

BOX 1-1
Converting Biomass to Ethanol

Plant cell walls are the most abundant form of biomass on the earth and thus an immense potential carbon source for biobased products. Recent advances in biotechnology may now make it possible to exploit this raw material for the production of valuable commodities such as ethanol.

Plant cell walls are composed of crystalline bundles of cellulose embedded in a covalently linked matrix of hemicellulose and lignin. This complex polymeric structure poses a formidable challenge for solubilization and bioconversion. Dilute acid can hydrolyze (break down) hemicellulose at 140°C to yield pentose (5-carbon) and hexose (6-carbon) sugars. These simple sugars (predominantly xylose and arabinose with some glucose) are common substrates for bacterial metabolism. However, no naturally occurring organisms yet cultured can rapidly and efficiently convert both pentoses and hexoses into a single product of value.

Advances in genetic engineering have made it feasible to redirect the metabolism of simple sugars in certain bacteria so that they form no unwanted byproducts and efficiently channel key metabolites only into a desired end product. This approach was initially taken by Lonnie Ingram and colleagues at the University of Florida to create a strain of common bacterium, *Escherichia coli*, having an altered metabolism that diverts carbon flow to ethanol. The scientists inserted genes cloned from *Zymononas mobilis* into the chromosome of *E. coli*. These genes code for the enzymes pyruvate decarboxylase (which converts the intermediate pyruvate into ethanol) and alcohol dehydrogenase (which makes the conversion more efficient). Pyruvate decarboxylase binds pyruvate more tightly than the enzyme lactate dehydrogenase which, in unaltered *E. coli*, converts pyruvate to lactate.

Although seemingly straightforward, this experiment in metabolic engineering was based on a great deal of genetic, biological, and biochemical information resulting from years of effort by many researchers. Previous work had shown that the *E. coli* chosen for the "production" strain could metabolize all of the major sugar constituents of plant biomass. More important, tools for the genetic and biochemical manipulation of *E. coli* were available only because the bacterium had been subject to intense study, making it perhaps the best characterized of all bacteria.

Ingram and his colleagues have now gone beyond their initial work with *E. coli* to engineer the cloned *Z. mobilis* genes into other bacteria such as *Klebsiella oxytocia* and *Erwinia* species. Unlike *E. coli*, these bacteria require less preliminary treatment of the cell walls because they contain additional enzymes that allow direct uptake of complex sugars (such as cellobiose and cellotriose) from plant cell walls. *Erwinia* strains also contain enzymes called endoglucanases that aid in the solubilization of lignocellulose.

The scientists' success in engineering bacteria to produce a valuable commodity like ethanol is the first step. Further work is now needed to make these processes economically competitive with production of ethanol from petroleum-based materials.

SOURCE: Beall et al. (1991); Beall et al. (1992); Ingram and Conway (1988); Ingram et al. (1987); Wood and Ingram (1992); Mohagheghi et al. (1998).

carbon dioxide are complex, and this topic continues to be an active area of research (see, for example, Marland and Schlamadinger, 1995).

The production of feedstocks for biobased industries could pose some problems to the environment, and these potential problems should be evaluated and minimized. There is the potential to use marginal land to grow crops that pose low risk of soil erosion or loss of wildlife. Impacts will depend on a number of factors, such as previous use of land, the planted crop, and crop management practices (OTA, 1993). Indiscriminate production of grain or removal of crop residues on vulnerable land could enhance erosion, degrade soil quality, increase flow of sediments and nutrients into surface waters, encourage herbicide use, and damage various ecosystems. Production of perennial grasses or woody crops and minimization of agrochemical inputs could limit such impacts. The use of perennial grasses and woody crops, moreover, could have environmental benefits by reducing erosion and improving soil structure and organic content as well as water quality (Hohenstein and Wright, 1994). Widespread impacts of harvesting residues on soil quality are not well understood, but some some research indicates that an estimated 80 million metric tons[1] of crop residues might be removed without impacting soil conservation measures (OTA, 1980). However, excessive amounts of crop residue should not be removed from farmland so that the residue can continue to build soil organic-matter levels. Conversely, harvesting residues for production of biobased chemicals may reduce air pollution from the open burning of residues and the frequency of plant pest and disease outbreaks, thereby reducing fungicide and insecticide use. At the same time research is done that will lead to more economic conversion of agricultural wastes, analyses should be done to examine the consequences of large-scale diversion of agricultural wastes for use as feedstocks for biobased industrial products.

Life-cycle assessment has emerged as a valuable decision support tool for both policymakers and industry in assessing the cradle-to-grave impacts of a product or process. The International Organization of Standardization is developing standards based on life-cycle analysis methodology for wood-based and other products. The significance of life-cycle analysis is underlined by the 1993 executive order by President Clinton requiring life-cycle analysis for federal procurement of environmentally preferred products. Such analyses should be holistic and include environmental and energy audits of the entire product life cycle, rather than a single manufacturing step or environmental emission. While the envi-

---

[1] The term *ton* as used in this report refers to metric tons

ronmental consequences of biobased production are expected to be largely positive to neutral, assessment of environmental impacts from biobased products should be continued.

Rural Employment

Farmers and rural communities could benefit from the employment and business opportunities that would result from production of biobased industrial products, by either growing new raw materials or providing locations for processing plants. Biobased industries will probably be sited near where feedstocks are grown in order to reduce transportation costs. Thus, industrial opportunities for biobased products would tend to appear throughout agriculturally productive areas of the country. While there may be some potential for biobased industries to increase job opportunities, there are insufficient data to make accurate predictions of the impacts of biobased industries on future employment trends.

Currently the entire chemicals industry (not the fuels industry) employs roughly 1 million people with annual sales of about $250 billion dollars. A ratio of labor employed to annual sales will yield a multiplier of about $250,000 in annual sales per job. An Economic Research Service study on the crambe industry (ERS, 1997b) showed $10 million in total sales giving 42 new jobs, which is almost the same ratio, $250,000 in annual sales per job. Considering the multiplier effect, for every primary job in manufacturing, approximately four new jobs are created in service and supplier industries. Ultimately, there would be a lot of processing plants, and this committee can envision around a million jobs based on processing agricultural and forest raw materials to chemicals only, without taking such fuels as ethanol into account. However, new employment opportunities provided by the biobased industry would to some extent be offset by decreases in employment in the petrochemical industries. This is a topic that warrants further research.

Diversification of Petroleum Feedstocks

Current and potential oil reserves are substantial, and exploration continues to open new petroleum supplies for the world market (eg., Caspian Sea). There does, however, remain an open question as to the size of petroleum reserves and the future cost of petroleum products. Experts estimate that two-thirds of the world's proven reserves are located in a single geographic region, the Persian Gulf, and this area will continue to serve as a dominant source for oil exports (USDOE, 1998). However, some geologists report that oil reserves could be depleted in only 20 years (Kerr, 1998). According to the American Petroleum Institute, there were approximately 43 years' worth of reserves remaining as of 1997, an increase from the 34 years prevailing before the first OPEC crisis of 1973. As a substitute for oil, biomass could help diversify feedstock sources that support the nation's industrial base. Policymakers should consider the potential economic impacts from large-scale biobased fuel production on world energy prices. In the near term, biomass feedstocks could help minimize price and supply disruptions in international petroleum markets. However, introduction of massive quantities of energy substitutes on the world market could lead to falling oil prices, creating a larger gap between the prices for biobased and petroleum-based industrial feedstocks. While this committee cannot predict with any accuracy the availability and cost of future supplies of petroleum, the committee believes that the United States can lead the transition to greater use of renewable materials as oil and other fossil fuels are gradually depleted.

SETTING A COURSE FOR THE FUTURE

Many recent technical and economic assessments show that the United States has the potential to return to a carbon economy based on renewable biological resources (ERS, 1990; ERS, 1993; Harsch, 1992). Growing public concerns about pollution and the environment have intensified interest in new uses for agricultural and forestry resources. Biobased industries may provide farmers with new markets beyond the traditional food, feed, and fiber products. Recent advances in the biological and materials sciences are leading to the development of new and less costly technologies for growing and processing plant matter and for manufacturing biobased products. Many opportunities are on the horizon for biobased industrial products, and both public and private interest have been sparked. New chemicals and materials isolated or manufactured from renewable resources promise industrial products with superior performance characteristics (Kaplan et al., 1992). The future of a biobased industry depends on products that outperform petroleum-based products at a competitive cost.

Much more work is needed to realize the full promise of biobased products. Both federal and private research funding in this area has been sporadic over the past decade. The development of new or improved processing technologies will largely determine which biobased products become available. While certain processing technologies are well established, others show promise but will require additional refinement or research before they come into practical use. The committee believes that the potential benefits derived from biobased industrial products could justify public policies that encourage a transition to renewable resources.

REPORT COVERAGE

This chapter has examined the significance of carbon in the economy and identified potential consequences of relying on fossil versus biological sources of carbon. An increased emphasis on biobased industrial products could enhance access to diverse markets, provide environmental advantages, and diversify sources of strategic feedstocks. Whether such a shift occurs will depend on public policy decisions and developments in several key areas addressed in the remainder of this report.

Chapter 2 examines existing and potential renewable raw materials that could be used as a source for biobased industrial products. The chapter provides an overview of current production of plant materials and describes the potential for increasing the variety and amounts of plant material available for industrial uses. It also addresses applications of technologies to develop new resources such as genetically modified plants and microorganisms.

Chapter 3 considers some of the most significant current examples of biobased industrial fuels, chemicals, and materials. An outline of the scope, magnitude, and developmental dynamics of such products is presented to provide a framework for analyzing future prospects.

In Chapter 4 the committee discusses biomass processing, covering thermal, mechanical, chemical, and biological processes. The chapter focuses in particular on the development of biorefineries as an essential step for biobased industrial products to replace most fossil-based products.

Chapter 5 presents the committee's major conclusions and recommendations derived from analyses in the preceding chapters. Here, the committee describes opportunities to integrate science and engineering to reduce the cost of processing abundant raw materials into value-added biobased products. The chapter identifies specific priorities for investment in research, development, and commercialization and summarizes the public- and private-sector activities that could accelerate the growth of a biobased industry in the United States.

2

Raw Material Resource Base

The United States has abundant forests and croplands, favorable climates, accessible capital, and sophisticated technologies for a strong biobased industry. As agriculture productivity and silviculture productivity continue to increase, more biomass will be available to support a biobased industry. Advances in biotechnology will keep a continuous supply of new crops flowing into the marketplace. The United States has substantial resources to invest in a carbon economy based on renewable resources.

Conversion of industrial production to the use of renewable resources will require abundant and inexpensive raw materials. The three potential sources of such materials are agricultural and forest crops and biological wastes (e.g., wood residue or corn stover). The amount of each resource available for biobased production will depend on how much these crops are consumed by competing uses and how much land is dedicated to crops grown for industrial uses. The land and other agricultural resources of the United States are sufficient to satisfy current domestic and export demands for food, feed, and fiber and still produce ample raw materials for biobased industrial products except for massive fuel production.

SILVICULTURE CROPS

Forests are a major source of raw materials for the production of wood products. The amount of land supporting the nation's forests has remained relatively constant since 1930 (USDA, 1995). Heightened public interest for forest preservation has led to government policies that support conversion of federal forest lands to special uses such as parks and wildlife areas that prohibit timber production. As these competing uses for national federal forests intensify, increases in timber harvesting on private forestlands will have to offset timber production declines on U.S. public lands (NRC, 1997; USDA, 1995).

Productivity from silviculture and timber harvests has increased on forest lands. The average volume of standing timber per hectare (800 x 10^9 cubic feet in 1991) is now 30 percent greater than it was in 1952 (USDA, 1995). Forest growth nationally has exceeded harvest since the 1940s—a trend that was accelerating until very recently. In 1991 forest growth exceeded harvest by 22 percent, even though the harvest was 68 percent greater than in 1952. More recently (1986 to 1991), the proportion of timber harvested from the total forested land has increased, primarily as a result of increased harvesting on industrial forestlands. The U.S. Department of Agriculture (USDA) Forest Service is forecasting further increases in the nationwide volume of harvested timber from slightly over 16 billion cubic feet in 1991 to nearly 22 billion cubic feet in 2040 (USDA, 1995).

Production capacity of timbered forestland may be underused. Softwood residues are generally in high demand as feedstocks, but hardwood timber residues have less demand and fewer competing uses. Underutilized wood species include southern red oak, poplar, and various small-diameter hardwood species (USDA, 1995).

In the future forestlands may be planted to silviculture crops for use in bioenergy production. Bioenergy crops may confer a number of benefits such as low maintenance requirements, high yields, and environmental advantages. The USDA and the U.S. Department of Energy (DOE) have field tested several short-rotation woody crop species (harvested on a cycle of 3 to 10 years), including hybrid poplar, black locust, eucalyptus, silver maple, sweet gum, and sycamore. Certain woody feedstocks have yields averaging 4.5 to 7.5 dry tons per acre per year—two to three times the yields normally achieved by traditional forest management in the United States. Even higher yields occur under certain conditions. Recent results show potential yields that consistently reach 8.9 dry tons per acre per year in several locations (Bozell and Landucci, 1993).

AGRICULTURAL CROPS

Cropland acreage is the third major use of land in the United States. The most notable trend in cropland use is the movement of cropland from crop idling programs into crop use and out again (ERS, 1997a). Four principal crops—corn, wheat, soybean, and hay—accounted for nearly 80 percent of all crops harvested in 1996. Current use of commodity crops for industrial uses is low. Coproduction of grain crops such as corn for both food and ethanol fuels will help reduce any future conflicts inherent in allocating renewable resources to two important human needs: food and fuel.

The United States has long been the world's largest producer of coarse grains. Recent data indicate that domestic grain production makes up approximately 67 percent of the world's grain supply (USDA, 1997a). In 1998 the United States exported over 37 million metric tons of corn grain (ERS, 1999). According to the USDA, U.S. feed grain production is projected to increase steadily through 2005. Expected increases in production are due to increasing yields, except for corn, where more acreage also accounts for gains in some years. Corn yields are expected to increase 1.7 bushels per acre per year based on the long-term trend. Corn plantings are expected to remain at or above 80 million acres throughout the next decade (USDA, 1997a). Continuing gains in U.S. agricultural productivity will extend the resource base available for biobased crop production.

U.S. corn yield nearly tripled between 1950 and 1980 from an average of about 35 bushels per acre to over 100 bushels per acre (OTA, 1980). Based on these gains, analysts predicted that corn yields would exceed 120 bushels per acre by 1995. This has, in fact, occurred: the 1992 and 1994 corn yields exceeded 130 bushels per acre (NASS, 1994; USDA, 1997a). Of the 252 million metric tons of corn produced in the 1996 to 1997 marketing year, approximately 19 million metric tons (7 percent of the total corn grain production) were allocated to industrial uses (industrial starch, industrial alcohol, and fuel alcohol) (ERS, 1997b). To the extent that we understand the many factors contributing to crop yield, productivity increases in many cases will be enhanced by improvements in plant genetics, pest management, and soil quality. At the current rate of growth, another 19 million metric tons of corn could become available by the turn of the century.

Perennial grasses and legumes are being evaluated as potential energy crops (Hohenstein and Wright, 1994). These grasses include Bahia grass, Bermuda grass, eastern gama grass, reed canary grass, napiergrass, rye, Sudan grass, switchgrass, tall fescue, timothy, and weeping love grass. Legumes that have been tested include alfalfa, bird's-foot trefoil, crown vetch, flatpea, clover, and *Sericea lespedeza*. In 1992 about 150 million tons of hay (more than half of which was alfalfa and alfalfa mixtures) were harvested from 59 million acres of croplands in the United States (USDA, 1994). In 1994 hay was harvested from approximately 61 million acres in the United States (ERS, 1997a). Considerable preproduction research now focuses on the facile conversion of some of these materials into fermentable sugars. Thick-stemmed perennial grasses, such as energy cane and napiergrass, produce yields from 5.4 to 14.5 dry tons per acre per year. These are current yields and likely would increase following selection. They may one day be grown and used on a large scale.

ENHANCING THE SUPPLY OF BIOMASS

The amount of cropland that will actually be used to supply biobased processors depends on a demand for the final product, and the inputs used to make that product must be competitively priced. Industrial processors bid for corn and forages based on processing costs and product prices in the petrochemical and specialty chemical industries. Some industries that produce specialty starches and lactic acid plastics can bid grain and productive croplands away from food processors now. But some industrial products, such as grain-based ethanol, may not be able to compete with food producers even after considerable declines in grain or forage prices. Even with anticipated new technology, grain-based ethanol probably will not compete with petroleum fuels on a cost basis (Kane et al., 1989). Similarly, access to major commodity plastics markets, like ethylene, may require very low-cost feedstocks (Lipinsky, 1981). The amount of land devoted to crops for biobased industries will depend on economics, as tempered by agricultural policies.

Some resources that are not useful for food production may soon become more suitable for industrial products because processing technologies that use woody biomass are improving. Potential supplies from three sources—crop residues, wood wastes, and Conservation Reserve Program land—are discussed below. Other available biomass wastes (e.g., municipal solid waste) also may be potential sources of lignocellulosic materials. These reserves may provide the best odds for competitive production of biobased industrial products.

Waste Materials

The United States produces abundant wastes that are potential raw materials for biobased products. It is estimated that 280 million metric tons per year of biological wastes are currently available (refer to Table 2-1). Much of this is crop residues, predominantly from corn—about 100 million metric tons of corn residues are produced annually (Gallagher and Johnson, 1995). To a lesser extent, paper mill, wood, and municipal solid waste also are important. Approximately 5.6 million metric tons of unused wood residue is generated in all U.S. sawmills (Smith et al., 1994). Crop residues represent a major untapped source of carbon-rich raw materials available onsite at a low to negligible cost. However, expenses for collection, storage, and transport must be considered in using these bulky,

TABLE 2-1 Estimated Available Waste Biomass in the United States[a]

Feedstock	Quantity (1000 dry metric tons)
Recycled primary paper pulp sludge	3,400
Urban tree residue[b]	38,000
Mixed office paper	4,600
Sugarcane bagasse	700
Newsprint	11,200
Rice	2,700
Corn gluten feed	5,700
Spent brewers grains	1,100
Distillers' dried grains	1,800
Corn gluten meal	1,100
Small grain straw	0
Wood mill residue[c]	5,600
Corn stover[d]	100,000
Cotton gin waste[e]	15,000
Sulfite waste liquor[f]	61,000
Cheese whey from dairy[g]	28,000
Total	279,900

SOURCE: Rooney (1998).

[a] Available quantity estimates of various lignocellulosic resources as biobased feedstocks will vary with region and local soil conditions.

[b] Urban tree residue is a feedstock of heterogeneous quality available from many sources.

[c] Value of all unused wood mill residues from Smith et al. (1994).

[d] Corn stover estimate based on Gallagher and Johnson (1995). Calculations assume 30 percent residues left on soil. Some levels of corn stover may cost less than new crops because stover and grain are produced together making recovery of land costs unnecessary because they have already been accounted for in grain profit calculations.

[e] Cotton gin waste based on a density of 1 ton per acre. Data from personal communication with Ralph Hardy of the National Agricultural Biotechnology Council, September 21, 1998.

[f] Represents sulfite waste liquor generated in sulfite paper mills in the United States. Data from Morris and Ahmed (1992).

[g] Represents total amounts generated in sugar and cheese processing in 1990. Data from Morris and Ahmed (1992).

low-valued residues. Sufficient biological wastes exist to supply the carbon for all 100 million metric tons of organic carbon-based chemicals consumed annually in the United States as well as to provide part of the nation's fuel requirements (Morris and Ahmed, 1992). Production of industrial products from agricultural wastes can reduce competition for agricultural resources.

Conservation Reserve Program

There is potential that land idled by the Conservation Reserve Program (CRP) could be used to grow biobased crops. This federal program was initiated in 1986 to help owners and operators of highly erodible croplands conserve and improve the soil and water resources on their farms and ranches through long-term land retirement. The CRP provides monetary incentives for farmers to retire environmentally sensitive lands from crop production for 10 to 15 years and to convert them to perennial vegetation. The 1996 Federal Agricultural Improvement and Reform Act limited enrollment to 36.4 million acres through the year 2002 (ERS, 1997a).

Some CRP lands may be suitable for harvest of perennial grasses and energy crop production while preserving soil and wildlife habitat. Judicious harvesting on a fraction of CRP lands might be consistent with wildlife and wetlands preservation. Field-scale studies are under way to quantify changes in soil and water quality and native biodiversity due to production of biomass energy crops on former agricultural lands (Tolbert et al., 1997; Tolbert and Schiller, 1996). Grass production on CRP lands could enhance biomass supply: at least 46 million tons of additional feedstock would be available if one-half of CRP lands was available. This figure assumes low yields of biomass (approximately 2.5 tons per acre), and these yields could increase up to 10 tons per acre for some crops (e.g., switchgrass).

Land costs in the CRP are a barrier to the biobased industry. Land values are high because the federal government must at least match the opportunity of foregone profits from continued production of annual crops such as corn or wheat. Average rental costs under the 1997 CRP are about $40 per acre (Osborn, 1997). Without the CRP and with conservation requirements on these lands, biomass production might be competitive with lower land rental rates for pasture; comparable rental rates for midwestern pasture are about $20 per acre. If the rental rate for CRP lands fell to the pasture rate, the cost of producing switchgrass could decline; presently switchgrass production costs are about $40 per ton (Park, 1997). CRP revision for energy crop harvest is a contentious issue. Furthermore, some have argued that reduced land rental costs for energy crops are a de facto subsidy (Walsh et al., 1996). The potential of using CRP lands to grow biomass energy crops is a topic that merits further investigation.

The total biomass is sufficient to easily meet current demands for biobased organic chemicals and materials. High-value chemicals are not expected to require large acreages. Future demands for biobased commodity chemicals potentially can be met with biomass from waste resources and crops grown on some CRP lands. While biobased materials such as lumber, cotton, and wool do have substantial markets, these products now compete successfully for land resources.

Coproduction of human food and animal feed products such as protein with biobased products is expected to help prevent future conflicts between production of food and biobased fuels. Corn-based refineries, for example, yield protein for animal feed and oil, starch, fiber, and fuel alcohol products. In the case of pulp and paper mills, pulp, paper, lignin byproducts, and ethanol can be produced while recycling waste paper in a single system. Current demands for liquid fuel are being met with current production of corn grain. If policymakers chose to increase ethanol fuel production beyond the capacity for coproduction of food and liquid fuel, biobased crops grown for energy uses could compete for land with food production. Opportunities for coproduction of food, feed, liquid fuels, organic chemicals, and materials are described in more detail in Chapter 4.

FILLING THE RAW MATERIAL NEEDS OF A BIOBASED INDUSTRY

The foundation of a biobased industry depends on an abundant supply of plant materials. Raw materials such as starches, cellulose, and oil can already be extracted from plants for the production of biomaterials, chemicals, and fuels. The committee envisions that many more plant substances (e.g., biopolymers or chiral chemicals) may serve as raw materials for industrial applications in the future. While conventional breeding methods continue to play an important role in developing new crops and cultivars, genetic engineering of existing crops will greatly enhance the number and precision of such modifications and the variety of plant products available for industrial use. Introduction of new crops for biobased production will be limited without an adequate infrastructure for cultivar research, development, and commercialization.

Satisfying the raw material needs of expanding biobased industries will require crops with the following characteristics: contain biomolecules and biochemical systems with potential industrial applications; can be manipulated to produce desirable molecules; can sustain a high level of predictable raw material production; and are supported by an infrastructure for biomass harvesting, transfer, storage, and industrial processing.

Current Resources

Renewable resources have been used for a wide variety of industrial purposes. For example, siliviculture crops have been used for many years as construction materials; wood can undergo additional processing to yield a variety of other products such as paper and textiles. Because agricultural and silvicultural crops are highly variable, plant parts become more valuable when they can be further separated into their biochemical components. In the long term, plant parts that can be converted to sugars for fermentation are likely to become a major feedstock for the production of biobased chemicals, fuels, and materials.

Woody Plant Parts (Lignocellulosics)

Woody plant parts are an abundant biological resource for the biobased industry. Wood is a complex material composed of carbohydrate and lignin polymers that are chemically and physically intertwined. Considerable energy is required to separate the wood polymers from each other. Much of the harvested wood is used for lumber, and both wood and other woody materials are used for pulp production.

Improvements in wood processing are leading to new biomaterials that may replace plastic products currently produced from petrochemical sources. For instance, improved resistance to insects and decay fungi, dimensional stability, hardness, and other properties result when wood is esterified, cross-linked, and impregnated (Stamm, 1964). Wood that has been plasticized by decrystallization and esterification can be further shaped by injection- and extrusion-molding processes. Although these processes are currently economically impractical, they might provide many new products from wood-flour and wood-fiber-reinforced thermoplastic composites in the future. Scientists and engineers will continue to refine industrial processes that can be used to produce useful biomaterials from wood.

Separated Plant Components

Plant matter contains hundreds of components that are useful inputs for biobased production. Plants are primarily carbohydrates (cellulose, other polysaccharides such as starch, and sugars), lignins, proteins, and fats (oils). Starch and sugar polymers are end products of photosynthesis and dominate a plant's carbohydrate reserves. Accessing these vast carbohydrate reserves will be key to maintaining a renewable source of raw materials that can substitute for petrochemicals.

Cellulose

Cellulose is a carbohydrate polymer composed of glucose and constitutes about 45 percent of woody plant parts. Cellulose can be isolated by pulping processes and then further processed to yield such chemicals as ethanol and cellulose ethers; cellulose acetate, rayon, and cellulose nitrate; cellophane; and other cellulosics. Many of these derivatives have only specialty applications because their cost is high relative to that of petrochemical-derived polymers.

Numerous sources of cellulose pulp can be used for chemical production. The primary source of wood cellulose pulp comes from conifer species (Smith et al., 1994), but hardwood uses have increased in the past two decades. Flax residue (flax tow) and kenaf are grown commercially for pulp production. In other countries, pulp is made from crop residues such as straw and sugarcane bagasse. Because of its dominating abundance in plants, cellulose will always be a primary feedstock of any biobased industry.

Hemicelluloses

Hemicelluloses are composed of carbohydrates based on pentose sugars (mainly xylose) as well as hexoses (mainly glucose and mannose). Hemicelluoses make up 25 to 35 percent of the dry weight of wood and agricultural residues; they are second only to cellulose in abundance among carbohydrates. While use of hemicellulose is currently limited, quantities of hemicelluloses, pectins, and various other plant polymers are abundant in residues and have great potential in the production of chemicals and materials.

Lignin

Lignin is a phenylpropane polymer that holds together cellulose and hemicellulose components of woody plant matter. Lignin constitutes about 15 to 25 percent of the weight of lignocellulose. Lignin has not yet been used as a raw material for industrial use in large quantities. Concerted attempts by pulp and paper research laboratories to develop new markets for byproduct lignins have had only limited success (Bozell and Landucci, 1993). Production of low-molecular-weight compounds from kraft lignin (phenols in particular) similarly has not yet proved commercially competitive. This reflects the chemical complexity of lignin and its resistance to depolymerization. Nevertheless, a recent DOE study concluded that pyrolysis of lignocellulosics (lignin, cellulose, and hemicellulose plant tissues) could make production of phenolics and anthraquinone from lignin competitive, and the potential also exists to produce benzene, toluene, and xylenes from lignin via pyrolysis (Bozell and Landucci, 1993). Lignocellulose pretreatment receives special attention in this report because it will be a key step for realizing the presently untapped potential of abundant lignocellulosic materials found in wood and other perennial crops.

Starch

Starch is the principal carbohydrate reserve of plants. Corn starch currently is a primary feedstock for starch-based ethanol, plastics, loose-fill packing material, adhesives, and other industrial products. Approximately 600 million bushels of corn went into production of industrial products during the marketing year 1995 to 1996; of that total, 395 million bushels were used to produce fuel ethanol (ERS, 1996b). While the supply of corn starch has been sufficient to meet current demands, primarily anhydrous motor fuel grade and industrial ethanol, other supplies of sugar feedstocks are being evaluated to meet anticipated increases in demand for oxygenated fuels and chemicals.

Proteins

Proteins are the primary means of expressing the genetic information coded in DNA. These polymers are based on building blocks of amino acid monomers whose sequence is predetermined by a genetic template. The sequence diversity of proteins is responsible for the wide array of functions performed by proteins in living organisms (OTA, 1993) (see Box 2-1). A variety of plant proteins might one day be commercially exploited as materials, but current understanding of the structural properties of most plant proteins is limited.

One of the few well-understood plant proteins is zein, an abundant protein in corn seeds. Zein makes up 39 percent of the kernel protein, or about 4 percent of the kernel weight. The protein has several properties of industrial interest, such as the ability to form fibers and films that are tough, glossy, and grease and scuff resistant. Zein resists microbial attack and cures with formaldehyde to become essentially inert. In addition, it is water insoluble and thermoplastic.

The USDA Northern Regional Research Laboratory in Peoria, Illinois, developed zein into a textile fiber in the late 1940s. Scientists generated the fiber by dissolving zein in alkali, extruding the solution through spinnerets into an acid coagulating bath, and then curing the product with formaldehyde. Zein fibers are strong, washable, and dyeable and possess other desirable properties. The Virginia-Carolina Corporation commercialized zein-based fiber as "Vicara," producing about 5 million pounds in 1954. However, the company discontinued manufacture shortly thereafter, perhaps because of the advent of comparable synthetic fibers. Zein's main use today is as a water-impermeable coating for pharmaceutical tablets, nuts, and candies. It also functions as a cork binder for gaskets and bottle-cap liners, a binder in ink, a varnish, and a shellac substitute. The advantageous properties of zein suggest that its industrial usefulness merits reexamination (Wall and Paulis, 1978). There is potential to alter

> **BOX 2-1**
> **Nature's Nylons**
>
> Common orb-weaving spiders spin as many as seven different types of silk fiber, each critical to the spider's survival. The fibers are made of proteins that are high-molecular-weight linear polymers exhibiting a broad range of mechanical and physical properties. Some silks are strong, tough framework filaments that support the web. Others are elastic filaments that absorb the kinetic energy of insects striking the web. Accessory filaments are produced that wrap captured prey or provide cocoon material. Silk fiber production imposes high energy and material demands on arachnids. Consequently, spiders have evolved to produce protein fibers that are highly efficient structures and one of the best high-performance natural materials.
>
> Spiders use dragline filaments to control their movement in the wind and to form the primary load-bearing framework of the orb web. The combined strength and toughness of dragline filaments are unmatched by any existing man-made fiber: they have a tensile strength two to three times greater than steel and an elongation-to-break ratio approaching 30 percent. Dragline filaments offer an attractive benchmark for next-generation materials because of their exemplary physical and mechanical properties and also because they are processed at ambient temperatures from aqueous media.
>
> The fibers in dragline silk are made up of glycine- and alanine-rich linear proteins containing oligopeptide units that function like the hard and soft segment units of conventional man-made polymers. The unique properties and processibility of dragline silk proteins result from the length and distribution of the protein segments as well as the amino acid sequence of each segment. Recombinant DNA technology has now made possible the development of bioengineered analogs that theoretically will perform as well or better than natural dragline filaments.
>
> Several academic, government, and industrial laboratories in the United States are currently conducting research along these lines. The initial focus has been expression of protein polymers in microbial hosts, resulting in the successful production of several structural proteins in experimental quantities. Researchers also have prepared several fibers and films from de novo designs produced via fermentation. Early experience suggests that the economics of protein polymer production will need to improve before high-volume production and commercialization of these polymers become possible. Additional challenges will be determining how the constituent protein structures and biological spinning apparatus influence the transition from aqueous solution to insoluble fiber and finding ways to mimic the effects of spider anatomy and physiology at an industrial scale.
>
> ---
>
> SOURCE: Tirrell (1996).

the structural properties of zein by genetic engineering to produce novel characteristics.

Plant Oils

Many crops can serve as sources of plant oils; currently soybeans account for 75 percent of the vegetable oil produced in the United States.

Soybean crops are a major target of plant oil research. Approximately 305 million pounds of soybean oil were used in nonfood applications such as livestock feed and the manufacture of resins, plastics, paints, inks, and soaps in 1996 (ERS, 1997b). Fatty acids derived from soybean oil are being converted into surfactants, emulsifiers, and alkyd resins for paints. Soybean oil can be chemically transesterified to produce biodiesel (methyl esters). In the future biodegradable lubricants may be produced from soybean oil; genetically modified soybean varieties hold the promise of yielding lubricant products that outperform petroleum-based lubricants (ERS, 1997b). Additional oilseed crops, some yielding oils with unusual properties, could be grown in the United States (e.g., petroselenic acid from coriander oil) (see Box 2-2). Over the near term the volume of oil produced for such uses will remain small relative to petrochemical sources.

Fermentable Sugars

Fermentable sugars are by far the largest feedstock that might support a biobased chemicals industry in the United States. A wide range of fermentable sugars can be found in crops and wastes from agriculture and silviculture. Major feedstocks include corn, wheat, sorghum, potato, sugarbeet, and sugarcane; other sources include potato-processing residues, sugarbeet and cane molasses, and apple pomace (Polman, 1994). Sugars can be produced directly or derived from polysaccharides (such as cellulose and starch) and then, via microbial fermentation, used to produce a wide range of commodity and specialty chemicals. Existing commercial fermentations primarily utilize glucose (6-carbon sugar) to produce ethanol, acetic acid, amino acids, antibiotics, and other chemicals. Over the long term new sources of glucose will be required to meet the demands of a biobased industry. Growth of a biobased chemicals industry will depend on production of cellulose-rich crops, including those currently under production (e.g., corn and alfalfa) and others that presently are not grown commercially (e.g., switchgrass and hybrid poplar).

Significant increases in glucose reserves are available from lignocellulosic substances found in most plants, crop residues, and waste paper. Cellulose can be hydrolyzed by acid to glucose, although much of the glucose is destroyed during this process. The second most abundant sugar, found in hardwood and agricultural residues, is xylose derived from xylan hemicelluloses. Xylose is relatively easily recovered by acid or enzymatic hydrolysis but can be fermented to ethanol only by a few naturally occurring organisms or recombinant microbes. The practical sugar yield from lignocellulosics would increase significantly if commercial fermentations could utilize xylose (a 5-carbon sugar or pentose) as well as glucose (a 6-carbon sugar or hexose). Novel genetically engi-

> **BOX 2-2**
> **Evaluating Alternative Crop Sources of Petroselenic Acid**
>
> Coriander is grown primarily for its use as a spice, but it may have potential industrial uses. The plant's seeds are 17 to 20 percent triglyceride, 80 to 90 percent of which is (esterfied) petroselenic acid (an 18-carbon fatty acid with a single unsaturated bond at C-6). Oxidative cleavage of the petroselenic acid's double bond yields adipic acid, a 6-carbon dicarboxylic acid used in the synthetic polymers or nylon industry, and lauric acid, a 12-carbon fatty acid used in the soaps and detergent industry. Also, the derivation (hydrolysis) of the petroselenic acid ester from the seed triacylglyceride yields glycerol, another chemical with industrial utility. All of the carbon atoms in the seed oil can thus be used with no intrinsic processing losses.
>
> Coriander seed yields vary greatly (100 to 2,000 kilograms per hectare in Europe and India). Experimental yields in Europe have been as high as 2,500 kliograms per hectare. Production of coriander for industrial purposes would become an attractive area for research and commercialization if high yields can be maintained over large areas and if the processing chemistries for industrial feedstocks are made practicable on a greater scale.
>
> The desirability of developing coriander as a crop source of petroselenic acid needs to be weighed against other potential biological sources, however. Genetic engineering may one day produce a comparable source of this fatty acid in an agronomically adapted crop. Petroselenic acid is produced as an offshoot from the usual pathways of fatty acid triacylglycerol as petroselenic acid biosynthesis. The process involves desaturation of a 16-carbon fatty acid (palmitic acid bound to a protein) at C-4 by a specific desaturase, elongation by the usual fatty acid elongation reactions (although there may be additional requirements for a specific elongase activity), and cleavage from the protein by a thioesterase before incorporation into triacylglycerol as petroselenic acid.
>
> The key desaturase gene has been cloned and introduced into tobacco plants. Up to 5 percent of the total fatty acids in the modified tobacco's cells was petroselenic acid. Introducing the thioesterase and elongase genes along with the desaturation gene might yield greater quantities of petroselenic acid. Transgenic technologies might eventually produce seed oils with petroselenic acid levels rivaling that of coriander (up to 90 percent). If accomplished in a high-yielding oilseed crop, such as sunflower or rapeseed, the result would be a very high-yielding source for petroselenic acid that is stable, agronomically suited to the United States, and supported by a large agricultural and genetic infrastructure.
>
> ---
>
> SOURCE: Based on Dormann et al. (1994), Cahoon and Ohlrogge (1994a,b), Kleiman and Spencer (1982), Meier zu Beerentrup and Roebbelen (1987), and Ohlrogge (1994).

neered microorganisms will eventually play a key role in the direct conversion of cellulose oligomers and 5- and 6-carbon sugars to ethanol.

To avoid destruction of sugars from lignocellulosic materials by acid treatment, enzymatic hydrolysis using mixtures of enzymes (cellulases and hemicellulases) is used. These enzymes, when combined with effective pretreatments of lignocellulosics, provide high yields of glucose, xylose, and other fermentable sugars with minimal sugar losses. However, these enzymes are currently too costly to use in large-scale conversion of lignocellulosic materials to fermentation substrates.

Improving Plant Raw Materials

The discoveries occurring in plant and microbial genomics will advance the fundamental biological research needed to support a biobased industry. Scientific investigations are under way to decipher the genetic code of a flowering plant, *Arabidopsis thaliana*; the genetic map is complete for the microbial organisms *Saccharomyces cerevisiae* (common yeast), *Bacillus subtilis* (gram-positive bacteria), *Escherichia coli* (gram-negative bacteria), and *Caenorhabditis elegans* (nematode). In addition, it is expected that complete genomic sequences of Drosophila, humans, and several other eukaryotic species will become available in the foreseeable future. The genetic information collected on these organisms will provide researchers with insights on the genes that control plant traits and cellular processes (NRC, 1997). For example, understanding of the functions of the *Arabidopsis* genes will permit identification of a desirable gene, which on transfer to a different plant will have the gene's functions expressed there (e.g., the manufacture of a particular chemical).

Genetic engineering is perhaps the most significant development in plant biology in the past two decades. It has profound implications for understanding the fundamental processes of plant growth, development, and metabolism and for generating new agricultural and forest products. It is now possible to modify an organism genetically so that the modified plant or microbe produces greater quantities of a particular polymer. It is also possible to transfer an entire biological process into new organisms.

Biomass production can be improved by development of new cultivars and crops with enhanced agronomic traits. Researchers will need to use both traditional plant breeding and genetic engineering techniques to improve yield and pest resistance of traditional crops and new crops. Techniques like genetic fingerprinting and markers can be used to facilitate classical plant breeding. The identification of crop strains and the durability of genes that confer resistance to pests and environmental stress will contribute to enhanced productivity, as it already does with food and feed crops.

Significant screening efforts are needed to identify carbohydrates, lipids, and proteins with industrial potential. Development of commercial plant sources for these compounds through breeding and genetic engineering will require elucidation of the genes and enzymes responsible for the production of compounds in the source organism (plant,

> **BOX 2-3**
> **Genetic Engineering Methods**
>
> Traditional breeding is restricted to mobilization of genes within related plant species. In contrast, genetic engineering, through the process called transformation, allows scientists to transfer genes between not only unrelated species but also the kingdoms of living organisms. Transformation involves the introduction of DNA into plant cells and tissues. It changes the hereditary material in each cell of the altered plant, as well as the plant's biochemical reactions. Newly introduced traits might affect plant growth, development, nutrition requirements, nutrient content, or composition of harvested plant parts.
>
> Plant transformation is one of the fundamental tools by which genetic engineers modify plants. However, the techniques have only been developed over the past two decades. In the late 1970s, scientists discovered that the common bacterium *Agrobacterium tumefaciens* causes plant tumors when oncogenes are transferred from the bacterial Ti plasmid into plant chromosomes. Scientists at Monsanto Company and Washington State University, St. Louis, developed methods to delete the oncogenes and replace them with different genes of interest, thereby using the Ti plasmid as a vehicle to transfer desired genes into plant chromosomes. To ensure that all plant cells in an experimental mixture were transformed, they added a selectable marker gene (e.g., kanamycin resistance) to the transferred DNA (T-DNA). Exposure to selective growth conditions (e.g., a medium containing kanamycin) would then kill all of the nontransformed cells. Many plant cells are totipotent—an individual cell can grow into a whole plant. Thus, researchers could grow whole plants from individual transformed plant cells and select the plants that passed T-DNA to their progeny in a Mendelian-dominant manner. Various dicotyledon plants have been transformed using the *Agrobacterium* technology, including tomato, hybrid poplar, potato, soybean, cotton, rape, and sunflower.
>
> *Agrobacterium*-mediated transformation initially did not work for most monocotyledon plants, including the majority of grain crops (the exceptions are rice and banana). Various academic and industrial laboratories developed new technologies for monocotyledon transformation based on particle acceleration. "Biolistic" guns shoot DNA into plant cells; the cells incorporate the DNA into their chromosomes and recover. Electroporation involves putting cells into an electric field.

animal, or microbe). As our understanding of plant metabolism continues to improve, scientists will be able to manage more sophisticated manipulations of these systems to produce the desired biochemicals in the desired quantities. Separations of plant components for industrial uses can also be improved by genetic engineering.

Biochemical pathways and genes can be mobilized within plants to create new products based on molecules that originate from nonplant sources such as microorganisms. Further, biomolecules often can be modified to facilitate purification. Such capabilities have no parallel in

> When a current is passed through a plant cell, DNA (a charged molecule) can enter and be incorporated into the chromosome. "Biolistic" gun and electroporation methods have now been used to transform various monocot plants, such as corn, wheat, barley, rice, banana, and oats. More recently, even *Agrobacterium*-mediated transformation has been successful in monocot transformation. Most genetic engineering work has focused on introducing genes to the plant cell nucleus because chromosomes found there are passed on to progeny. However, technologies for transforming the DNA of plant cell organelles, such as chloroplasts and mitochondria, have recently been developed. The technologies are most advanced for tobacco, where an excellent chloroplast transformation system now exists. Chloroplast transformation has several advantages: because chloroplasts are maternally inherited, the T-DNA cannot be transferred to wild relatives via pollen; gene expression is high; and introduction of genes is site specific. The technology also makes possible the efficient expression of bacterial genes and operons in plants.
>
> Although transformation introduces genes into plant cells, gene expression determines whether and how the introduced genes alter the plant's traits. Genes contain three distinct elements. First, the *promoter region* is the molecular switch. It determines the timing, the tissue in which the gene is expressed, and the quantity of a gene's product. Second, the *coding region* produces messenger RNA that usually provides a template for protein production. It determines the nature of the trait, such as increased starch or resistance to pests. Finally, the *termination region* or *polyadenylation signal* terminates RNA production and increases RNA stability by adding polyadenylate residues. Failure of any of the three regions can impair gene expression. Moreover, it can be difficult to obtain expression of certain coding sequences, even with appropriate promoter and termination regions. The genes for Bt (*Bacillus thuringiensis*) toxins that confer insect resistance are an example: it was necessary to design synthetic genes to circumvent the problems with poor Bt gene expression in order to develop useful insect-resistant crops. Similarly, transcription of plant DNA and the processing and stability of mRNA in various plant species are far from being well understood. Research in this area will be critical for harnessing the full power of biotechnology.
>
> Source: Horsch et al. (1984; 1985); Jenes et al. (1993); McBride et al. (1993); and Perlak et al. (1990).

coal and petroleum-based feedstock systems and are a major potential advantage for biobased products.

Well-developed technologies do exist for transferring genes into plants using bacterial plasmids or particle acceleration (see Box 2-3). Many plant species have thus far been transformed. Nevertheless, more efficient methods for transformation and regeneration would improve development of commercial products. Most current transformation methods rely on the time- and resource-intensive process of producing large numbers of transgenics that are subsequently screened for gene activity and "true-to-type" plants. The pace of discovery and product introduction could be accelerated if transgenes could be inserted at specific sites in the plant genome that consistently yield the desired gene activity level and that have little or no effect on the plant's agronomic fitness.

Transformation of "elite" germplasm is proving to be another methodological difficulty in genetic engineering. The present approach is to transform varieties most amenable to the process and then transfer the transgenes to elite germplasm by plant breeding. This is an inefficient process because evaluation and selection phases of research and development are prolonged, creating delays in the commercialization phase. Direct transformation and site-specific gene insertion might be especially useful for perennial plants because cycle times for evaluation and breeding of these plants are long, laborious, and impractical.

The current battery of promoters for gene expression (see Box 2-3) will be insufficient to meet the sophisticated expression profiles of the future. For example, redirection of carbon from starches and oils to other biopolymer (e.g., polyhydroxy alkanoate) production may be desired in specific cells of seed tissue during specific stages of seed formation. Further, subsequent degradation of the biopolymer during germination may be essential to permit carbon use for seedling growth. Such "fine tuning" of plant metabolism will require an extensive set of promoters from which appropriate selections can be made.

Enhanced Productivity

Biobased research should support improvement of plant productivity. First, any advances in research to improve crops will be relevant to biobased industrial crops. Second, the modern techniques used here for dealing with insect pests, pathogens, weeds, and stress are mostly transferable to plant design. High-level plant productivity on a consistent basis will be essential for supplying biomaterials for industrial production. Genetic engineering can enhance plant productivity by the introduction of traits that reduce farm inputs (e.g., pesticides, fertilizers, water), increase farm productivity, or modify biochemical content. Plant biotechnology products such as NewLeaf potato and Roundup Ready soybean are the precursors of a new generation of products to be commercialized over the next decade. Most of these products have been designed to improve agricultural crop productivity for food and feed uses. It is anticipated that the next generation of crops will be introduced to a broader range of markets, including biobased industries.

Environmental impacts from the release of genetically modified organisms continues to be an area of concern. An earlier report of the National Research Council concluded that crops modified by molecular and cellular methods should pose no risks different from those modified by classical genetic methods for similar traits (NRC, 1989). A finding of relevance is that established confinement options are as applicable to field introductions of plants modified by molecular and cellular methods as to introductions of plants modified by classical genetic methods.

Resistance to Insect Pests

Biotechnology companies are commercializing transgenic seed with resistance to insects in some major agricultural crops. Current approaches to insect resistance are based on expression of genes from the bacterium *Bacillus thuringiensis* (Bt) in plants. Bt genes produce toxins that can control caterpillars and certain families of beetles and mosquitoes. Bt genes are target specific and not efficacious against other major pests that cause significant yield losses. Additional genes will therefore be required to successfully manage the range of insects that are crop pests or that transmit plant viruses (e.g., whitefly, aphids). Moreover, pests continuously evolve, and the current generation of genes cannot provide the spectrum and durability of resistance required over the long term (NRC, 1996a).

Resistance to Plant Pathogens

Plant breeding will continue to be a predominant tool of defense against many plant diseases. While classical breeding techniques are fundamental to disease management, genetic engineering is becoming increasingly important. The first example of genetic engineering for disease resistance involved a gene encoding the coat protein of the tobacco mosaic virus (TMV) that was introduced into tobacco plants through an Agrobacterium vector (Abel et al., 1986). These plants were resistant to TMV as well as other closely related viruses. Expression of coat protein genes has become increasingly important in developing new varieties of resistant crops. Research is under way to transfer viral-coat proteins into other horticultural and field crops (Fitchen and Beachy, 1993). In 1995 a transgenic squash seed variety conferring this resistance characteristic was commercialized, and virus-resistant papaya, melon, tomato, potato, and other crops are in advanced stages of development. It is anticipated that transgenic techniques will enlarge the pool of disease-resistant genes that can be introduced into susceptible crop varieties (NRC, 1996a).

Managing plant fungal disease has attracted significant attention from biotechnologists, but progress has been slow. Bacterial chitinase reportedly confers resistance to *Botrytis*, a major fruit tree pathogen. Researchers claim that introduction of the resveratrol pathway into tobacco confers *Rhizoctonia* and *Botrytis* resistance (Hain et al., 1993). Several antifungal proteins have been identified in plant roots and seeds (AFP1 and AFP2 from radish and osmotin from tobacco), and their expression appears to delay infection by fungi (Terras et al., 1992). Whether the resistance demonstrated experimentally will be commercially useful remains to be seen. Based on the successes with insect and viral-vector control, the major missing link is identifying several efficacious antifungal genes (either proteinaceous or nonprotein) that can be used concurrently.

Researchers have identified some genes responsible for disease resistance in plants (Staskawicz et al., 1995). Natural disease-resistant genes generally produce proteins that recognize a ligand produced by an invading pathogen, causing cells exposed to the pathogen to undergo "hypersensitive response" (HR) and cell death. This process physically contains the pathogens and also produces one or more signals that activate the whole plant's defense system. This "systematic acquired resistance" (SAR) apparently confers broad protection against diverse pathogens. The mechanisms of HR and SAR are viable targets for engineering disease resistance into plants. The specificity of R genes to specific ligands, however, makes it difficult to confer multirace resistance to pathogens using these genes unless R genes can be identified that recognize certain basic ligands present in and essential to the pathogen. An alternative approach would be to gain an understanding of the HR and SAR mechanisms and introduce these "activation" mechanisms by way of other systems. Sophisticated promoters that can turn the HR or SAR processes "on" might be essential in these efforts.

Weed Control

Weeds cause major losses in crop productivity and lower crop quality. Introduction of conservation-tillage, reduced-tillage, and no-tillage practices to lower soil erosion have been accompanied by use of broad-spectrum herbicides to control weeds previously managed by mechanical means. It is expected that in the near term weed management will be dominated by pesticides and by use of genetically engineered or classically bred crop varieties. Research leading to development of new herbicides and transgenic crops will involve efforts of scientists from industry and academe. Several herbicides with low risks to human health and the environment have been developed (e.g., glyphosate, sulfonylureas, imidazolinones, and glufosinate; Kishore and Shah, 1988).

Some low-toxicity herbicides can only be used safely on a narrow spectrum of crops, thus limiting their utility. Researchers have successfully introduced resistance genes into crop varieties to protect crops against herbicides. For example, sulfonylurea-tolerant soybean, imidazolinone-resistant corn, and glyphosate-resistant soybean varieties have been commercialized.

Resistance to Environmental Stress

In addition to pests, various abiotic agents lower plant productivity. Extremes in temperature, water, and salinity create plant stress and lead to declines in plant growth, productivity, and quality. In the 1980s, drought caused corn yield losses and led to contamination with aflatoxins in many parts of the United States. High soil moisture was responsible for major crop losses in the Midwest in 1993. Knowledge of the genes conferring resistance to abiotic stresses is improving, and further research can be expected to minimize environmentally induced yield losses. Recent research has identified certain genes that might mitigate these impacts, such as the biosynthetic genes for fatty acids, sugars, and amino acids (Yoshida et al., 1995; Tarczynski et al., 1993). Some genes may be introduced into plants via conventional or molecular techniques to increase a plant's tolerance to various stress conditions. Other genes might reduce costs associated with weather-related crop damage and result in yield gains. For example, fertilizer inputs may be reduced by growing plants that make more efficient use of nitrogen, phosphorus, and sulfur or that can fix atmospheric nitrogen. Development of such cultivars could potentially reduce nitrate contamination of groundwater caused by fertilizer seepage. A fundamental understanding of plant nitrogen metabolism will be essential to identify genes controlling plant utilization of nitrate and nitrogen. Researchers have recently cloned some important genes (e.g., glutamine synthetase, asparagine synthetase, glutamate-oxoglutarate aminotransferase, nitrate reductase, nitrate and ammonia carriers) and are now beginning to elucidate their ectopic expression on plant growth and metabolism (Tsai and Corruzzi, 1993).

Plant Design

Research to enhance productivity, improve processing characteristics, and reduce the time required for harvest could lead to crops designed specifically for industrial applications. For example, a large quantity of biomass is left on the field in the form of plant residues. In the future it may be desirable to redesign plants to maximize harvested plant biomass for industrial processing. If dwarf corn plants could produce two to three ears instead of one or two, harvested biomass and grain yields might double from 150 to 250 bushels per acre up to 300 to 500 bushels per acre. While this scenario is unlikely in the near term, such yield enhancement could lower the costs of biobased production and enhance the competitiveness of biobased industrial raw materials. Conversely, corn fiber might be improved for industrial uses, in which case larger plants, rather than dwarf plants, might be desired.

Altered Biochemical Content

Genetic engineering allows scientists to manipulate the biochemical content of plants in unprecedented ways. The ability to access and alter the expression of biochemicals in seeds and other harvested components will be of particular importance to biobased crop research. The majority of grain crops do produce diverse carbohydrates, oils, and proteins. To exploit this chemical diversity, scientists will need to gain a more detailed knowledge of the plant genes and enzymes regulating these biochemicals.

Carbohydrates

Significant progress has occurred in altering the carbohydrate chemistry of agricultural crops. For example, potatoes containing novel carbohydrates have been commercialized and similar technologies are being applied to corn and wheat (Kishore and Somerville, 1993). While potatoes produce less starch than grain crops on a per-acre basis, starch-enhanced potatoes are being commercialized in some value-added markets such as food processing (see Box 2-4). Potatoes producing amylose-free starch are providing a desirable starch source for some markets because these modified potatoes contain only amylopectin rather than a mixture of amylose and amylopectin. Potatoes capable of fructan production will yield a new polymer based on fructose instead of glucose. Microbes can metabolize fructose and industrial processes can convert fructose to a wide range of organic compounds. Fructose is an example of the many sugars that are becoming important feedstock to the chemical industry.

Other polysaccharides such as cellulose, pullulan, hyaluronic acid, guavan, and xylans have interesting material, polymer, and fiber properties. The genes and enzymes involved in the biosynthesis of specific molecules will need to be identified and expressed in their natural hosts or engineered into other organisms. In some cases, genes may be modified to alter enzyme substrate specificity and accumulation of new polymers with different ionic charge, chemical reactivity and stability, solubility, melting, and other thermoplastic properties. The availability of genetic mutants in *Arabidopsis*, corn, and other plants will accelerate this research. A complete understanding of plant carbohydrate metabolism in unmodified and modified plant tissues will promote application of these sophisticated engineering technologies to other applications.

Lipids

Researchers have identified several plant genes and enzymes involved in lipid metabolism over the past five years, enabling development of modified oilseed rape cultivars for the biotechnology industry. Because of the similarity in lipid metabolism across plant species, similar oil research is under way with soybean, sunflower, and corn crops. Genes have been identified that affect carbon chain length, degree of unsaturation, and substituents in the fatty acid hydrocarbon chain (Topfer et al., 1995). Numerous fatty acids have been identified by screening the composition of various plant species. These plant species will provide a gene source for creating cultivars of major agronomic crops that produce fats and oils of value to the chemical industry as lubricants, fuels, and detergents. Introduction of additional functionalities, such as hydroxy or epoxy groups or double- and triple-carbon bonds, into plant fatty acids will enable synthesis of new molecules in major oilseed crops.

Many benefits may be derived from research that improves productivity and biochemical characteristics of some oil-producing crops. For example, the ricinoleic acid present in castor bean oil can be used for the production of nylon 11; the erucic acid found in crambe and rapeseed oils can be used for nylon 13 production; and the petroselenic acid present in coriander can be used for the production of nylon 66. High-linolenic oils present in the seed can be used to produce various coatings, drying agents, and printing inks. Significant opportunities may exist to improve the agronomic productivity of some of these oil crops and develop applications for the fatty acids and byproducts.

The oil pathway of plants may serve as a platform for the production of novel biopolymers. Industrial scientists have recently succeeded in transferring genes from the bacterium *Alcaligenes eutrophus* into the plant *Arabidopsis*. This modification led to plant production of poly(hydroxybutyrate) (PHB); this is an example of biopolymer engineering that can be performed on plants (Poirier et al., 1995). PHB constitutes nearly 20 percent of leaf dry matter in the genetically engineered *Arabidopsis*. Researchers at Zeneca and Monsanto are now transferring PHB genes into oilseed rape and soybean seeds for production of the polymer. Scientists anticipate that cotton fiber quality could be improved if PHB genes could be introduced into cotton plants. Future work will expand beyond PHB and focus on the production of diverse polymers that vary in carbon chain length and substitution. This work will require a more detailed understanding of fatty acid metabolism and the microbial pathways involved in polymer formation. Related research leading to identification of inexpensive processes for the extraction and separation of these polymers also will be critical for developing industrial applications.

> **BOX 2-4**
> **Genetic Engineering to Increase Starch Biosynthesis**
>
> Starch is the main storage carbohydrate in most plants. It is a major harvest component of several crops and thus directly affects yield. Industrial starch demand has increased dramatically over the past decade, primarily because of growth in the production of high-fructose corn syrups and bio-ethanol. In addition, various specialty starches, such as amylose and waxy starch, are being recognized for their superior material and nutritional properties as well as biodegradability.
>
> Plant breeders have worked to increase the starch content of potatoes for a number of years; however, breeders typically do not achieve large changes. When large changes do occur, they prove difficult to work with since they can involve multiple genetic loci. The genetic engineering work described here involved a single gene and dramatically increased starch production, and thus harvested biomass per acre, without any apparent harmful effects on the plant.
>
> The high-starch potato is one of the first genetic engineering products that targets industrial needs. Dry matter content (mostly starch) is the most important characteristic of potatoes for the processing industry because starch content affects processing cost, efficiency, and yield. For the food industry, high-starch potatoes are expected to improve efficiency and consume less oil during frying, thereby yielding fried products with more potato flavor, improved texture, reduced calories, and less greasy taste.
>
> The strategy taken by Monsanto researchers was to increase starch content by enhancing the rate of starch biosynthesis. Starch biosynthesis in plants (and glycogen biosynthesis in bacteria) requires the enzymes ADPglucose pyrophosphorylase (ADPGPP), starch synthase, and branching enzyme. These enzymes

Proteins

Little genetic engineering research has focused on proteins other than enzymes, although there are several advantages to using protein polymers for biobased production:

- Plant proteins generally are more diverse than other plant polymers.
- Molecular weight and amino acid sequences of protein polymers can be precisely regulated.
- Proteins can catalyze reactions or be hydrophobic, hydrophilic, neutral, acidic, or basic reactants.
- Proteins can form higher-order structures such as multimers of polymers.
- Plant proteins are generally inexpensive.
- Certain proteins, such as silk and wool, have long histories in the textiles industry.

> build the large starch molecule. The ADPGPP catalyzes formation of ADPglucose from glucose 1-phosphate and ATP. The ADPglucose subsequently serves as the substrate for starch synthase. Plant cells tightly regulate the activity of ADPGPP, turning the enzyme "on" when excess carbohydrates are present and shutting it "off" when starch biosynthesis is not needed. The researchers reasoned that, since ADPGPP controls the amount of starch produced, addition of another enzyme that is not subject to control by the cell would cause more carbohydrate to flow into starch. Scientists at Michigan State University had utilized such an enzyme from the common enteric bacterium, *Escherichia coli*; this enzyme was previously discovered at the Pasteur Institute. The mutant enzyme GlgC16 causes *E. coli* to accumulate high levels of glycogen because the bacterial cells do not regulate its metabolic state (glycogen is similar to starch). The Monsanto group obtained the gene encoding GlgC16 designed it to be active only in tubers and transferred it into russet Burbank potato plants (the dominant potato variety in North America).
>
> In some cases, tubers from transgenic plants containing the GlgC16 gene contained on average 25 percent to 30 percent more starch than tubers lacking the gene. Extensive U.S. field testing has shown the high-starch trait is stable in a number of different potato-growing environments. Furthermore, the trait has had no negative effect on plant growth or tuber yield, but, since the tubers contain more starch, the harvested dry matter per acre is substantially increased. The starch molecule in the transgenic potatoes, while more abundant, is structurally unchanged—important because the molecule's structure determines its end-use properties. The starch enhancement technology has now been extended to tomato and canola.
>
> SOURCES: Leung et al. (1986), Stark et al. (1992).

Other Biochemicals

The biochemicals that may be manufactured by plants are not limited to carbohydrates, oils, and proteins. Rubber is an important hydrocarbon produced by certain plants, and genetic engineering could enhance the amount and quality of rubber from these sources. Worldwide demand for rubber is growing at a dramatic rate as automobiles are becoming more common in the emerging economies of Asia. The opportunity to meet this need from renewable resources is real and deserves attention. Thus, genetic engineering could significantly enhance the biochemical diversity of the plant world and address some major issues in plant-based industries.

Although this chapter focuses primarily on plant biotechnology, developments in microbial biotechnology will also be key to the expansion of biobased production. Research on microbial systems is addressing the processing of plant products as well as the handling of society's wastes. Increased understanding of metabolic control in microorganisms is point-

TABLE 2-2 Crops with Potential Uses for Industrial Products

Crop	Current Acreage	Primary Products	Uses	Comments
Castor	Unknown (Major toxin-free seed is being accumulated in the U.S.)[a]	Ricinoleic acid (Castor seeds contain over 50% oil on a dry-weight basis and almost 90% of the oil is ricinoleic acid)[a]	Lubricants, plastics, coatings, sealants	Because of widely fluctuating world supplies and the structure of the world market, prices for castor oil vary considerably. This affects cash flow, makes corporate planning difficult, and discourages investment in new products. Commercial production of transgenic canola containing 15% ricinoleic acid is currently under way.
Crambe	35,000–40,000[b]	Erucic acid	Lubricants, waxes, paints, nylon 1313	Major competition by industrial rapeseed. Seed shattering and agronomy are major issues.
Cuphea	Unknown	Capric and lauric acids	Soaps, detergents, lubricants	A potential substitute for tropical oil.
Guayule	Unknown	Natural rubber	Rubber products, tires, surgical gloves, nonallergenic rubber products	Need to increase rubber content and utilize more of the plant.
Jojoba	16,000[c]	Wax esters	Cosmetics, lubricants, waxes	Needs more research into processing and use.
Kenaf	8,000[d]	Short and long fibers	Rope, newsprint, paper products	Potentially meant for replacing newsprint; most well-studied alternative crop.

Meadowfoam	5,000	High-value oils	Personal care products	Other uses under development; unsaturated chemical bonds confer unusually stable oil product.
Milkweed	<100[e]	Floss	Down quilts, pillows	Possibly large potential if nonwoven and yarn markets materialize.
Herbaceous crops (switchgrass, etc.)	600,000[f]	Biomass	Fermentable sugars, liquid fuels	30 to 40 million acres may supply up to 25% of U.S. liquid fuel needs.
Silviculture crops (short rotation, woody species such as hybrid poplar)	130,500[g]	Biomass	Fermentable sugars, liquid fuels, chemicals	30 to 40 million acres may supply 10 to 20% of U.S. liquid fuel needs.

[a] During the 1950s and 1960s, approximately 85,000 acres of castor were grown annually in the United States. Since then domestic production decreased and was abandoned in 1972 due to disagreement on an annual contract between castor seed processors and castor oil buyers. Source: USDA (1992).

[b] Source: Personal communication with John Gardner, Agro-Oils, Inc., Carrington, North Dakota, July 20, 1998.

[c] Source: USDA (1992).

[d] Estimated value. Source: ERS (1997b).

[e] Source: Personal communication with Herbert Knudsen, Natural Fibers Corp., Ogallala, Nebraska, July 17, 1998.

[f] Switchgrass is currently used by the livestock industry primarily for summer grazing in the Midwest and Great Plains. Source: Personal communication with Ken Vogel, U. S. Department of Agriculture, Agriculture Research Service, September 25, 1998.

[g] There are no hybrid poplar crops grown specifically for energy uses. Approximately 62,500 acres of hybrid poplar are planted in the Pacific Northwest; 54,000 poplars, sycamores, and sweetgum are planted in the Southeast; and 14,000 acres of hybrid poplar are planted in the North Central region for production of paper, boards, etc. To a limited extent, some of the residuals are being used for energy if the trees are planted very close to the mill. Source: Personal communication with Lynn Wright, U.S. Department of Energy, Bioenergy Feedstock Development Program, Oak Ridge National Laboratory, September 24, 1998.

ing to new ways to genetically modify organisms for industrial conversion of plant-derived feedstocks.

Introduction of New Crops

Most of the agricultural research in the United States focuses on major agronomic crops and developing applications for existing biochemicals extracted from these crops. Although this work is worthwhile and should continue, it will be equally important to develop new crops that have the potential to produce desired biochemicals. In addition to selecting and designing crops to meet certain industrial chemical needs, enhanced productivity should be a major goal of crop development efforts. This section briefly explores issues and opportunities associated with the introduction of new crops. Other analyses have described potential markets and botanical details of alternative crops (e.g., Harsch, 1992). Table 2-2 summarizes data for some new crops that have received initial scientific and commercial investments.

Basic technical factors often create difficulties in the commercialization of new crops. Many alternative crops lack characteristics that would result in high yields because they have been neither intensively cultivated nor subject to research and development for improved agronomic traits. Well-tended experimental plots may demonstrate useful genetic potential. However, the absence of directed plant breeding and underlying scientific knowledge generally makes for a long development period prior to major production of a new crop. In contrast the value of traditional crops is continuously being enhanced by technological advances resulting from major investments in research and development.

Biobased crops should be selected based on their productivity of the desired product as well as specific biochemical characteristics. Plant breeding may be necessary to introduce new biomolecules or enhance total biomass production. Sugar cane provides a useful example. Cultivars of sugar cane, called "energy cane," have been developed for ultimate conversion of the sucrose, cellulose, and hemicellulose contents to ethanol. Plant breeders developed cultivars having lower sucrose contents because the higher biomass yields of the cultivars more than compensated for the lower sucrose levels, thereby reducing the raw material's final cost.

Plants of natural origin and genetically engineered crops should be considered during the crop selection process. Castor provides an instructive example. The large-scale reintroduction of this crop is largely driven by a desire to replace the $30 million annual importation of castor oil with a reliable, cost-effective, domestic supply of ricinoleic acid. However, crops that produce high levels of oleic acid, such as sunflower or rapeseed, are being engineered to contain the gene required to produce hydroxyleic acid, thereby yielding the desired ricinoleic acid in an established agronomic crop.

Over the near term the acreage of traditional crops will continue to dwarf that of new crops. In the long-term, alternative crops can make important contributions in the industrial and agricultural sectors—if they can compete in the marketplace with traditional crops. Industrial crops that will be successful will be those with sufficient registered crop protection chemicals, appropriate infrastructure, optimized manufacturing processes and equipment, and byproduct utilization systems.

SUMMARY

If appropriate and sufficiently low-cost processing technologies were developed, there is enough unused biomass to satisfy all domestic demand for organic chemicals that can be made from biological resources (approximately 100 million tons per year) and all of the nation's oxygenated fuel requirements (use of oxygenated gasoline) in areas that did not meet the federal ambient air standard for carbon monoxide as mandated by the Clean Air Act Amendments of 1990. Production of biobased crops on land presently idled could, given low-cost technologies for converting these crops, provide an additional source of U.S. liquid fuels. A few new crops have received initial scientific and commercial investments, but various factors impede their commercial adoption. Nevertheless, certain nontraditional crops, such as switchgrass and hybrid poplar, are valuable because of their high yields.

Classical plant breeding and genetic engineering techniques will continue to be used by scientists for the development of new crops and improvement of well-established crops. Genetic engineering offers unprecedented opportunities to manipulate the biochemical content of specific plant tissues and design a raw material for easier processing—an advantage not enjoyed by fossil feedstocks. However, much more remains to be done to provide the raw materials for expanding biobased industries.

Over the long term, a major research priority is to maintain a commitment to fundamental and applied research in the biology, biochemistry, and genetics of plants and microorganisms. It is necessary to gain an understanding of underlying processes associated with gene expression, growth and development, and chemical metabolism. Improved methods of plant transformation and new promoters that further refine gene expression are needed to hasten the development of crops suitable for biobased industries. A sound scientific base in these fundamental areas will be critical to formulating strategies to supply future raw materials for biobased industries.

Future development of agricultural and forest crops for a biobased industry will strengthen the ties between agriculture and industrial production. The change will depend not only on continued improvement of traditional crops but also on the development of alternative crops, genetically engineered cultivars, and separation and fermentation processes that can make use of biomass. Making the transition to a competitive biobased industry will require close coordination between plant scientists and process engineers to develop cost-effective biological and industrial processes for the conversion of raw materials into value-added products.

3

Range of Biobased Products

At the turn of the century most nonfuel industrial products—dyes, inks, paints, medicines, chemicals, clothing, synthetic fibers, and plastics—were made from trees, vegetables, or crops. By the 1970s, organic chemicals derived from petroleum had largely replaced those derived from plant matter, capturing more than 95 percent of the markets previously held by products made from biological resources, and petroleum accounted for more than 70 percent of our fuels (Morris and Ahmed, 1992). However, recent developments are raising the prospects that many petrochemically derived products can be replaced with industrial materials processed from renewable resources. Scientists and engineers continue to make progress in research and development of technologies that bring down the real cost of processing plant matter into value-added products. Simultaneously, environmental concerns and legislation are intensifying the interest in agricultural and forestry resources as alternative feedstocks. Sustained growth of this burgeoning industry will depend on developing new markets and cost-competitive biobased industrial products.

Numerous opportunities are emerging to address industrial needs through the production and processing of biological materials. Today's biobased products include commodity and specialty chemicals, fuels, and materials. Some of these products result from the direct physical or chemical processing of biomass—cellulose, starch, oils, protein, lignin, and terpenes. Others are indirectly processed from carbohydrates by biotechnologies such as microbial and enzymatic processing. Table 3-1 shows

TABLE 3-1 Increase in Worldwide Sales of Biotechnology Products, 1983 and 1994[a]

	1983 ($ millions) [b]	1994[c] ($ millions)
Fuel and industrial ethanol	800[d]	1,500[e]
High-fructose syrups	1,600	3,100
Citric acid	500	900
Monosodium glutamate	600	800
Lysine	200	700
Enzymes[f]	400	1,000
Specialty chemicals[g]	1,300	3,000
Total	5,400	11,000

[a] Table excludes pharmaceutical products.
[b] Data from Hacking (1986).
[c] Data from John VicRoy, Michigan Biotechnology Institute, Market Analysis, 1994.
[d] Data based on Hacking (1986): 1983 ethanol price = $1.70 per gallon and volume = 180 million bushels corn (approximately 2.5 gallons ethanol per bushel). Total ethanol sales = $0.8 billion.
[e] Data based on February 21, 1994, Ethanol Profile, Chemical Marketing Reporter: 1994 industrial ethanol (fermentation) price = $1.70 per gallon and volume = 75 million gallons; fuel ethanol price = $1.1 per gallon and volume = 1.2 billion gallons. Total ethanol sales = $1.5 billion.
[f] Includes feed grades.
[g] Includes diverse products such as gums, vitamins, and flavors.

worldwide markets for several biobased industrial products (excluding pharmaceuticals) made from microbial and enzymatic conversion of carbohydrates. The gross annual sales of these biochemicals in 1994 exceeded $13 billion (Datta, 1994). Analyses of historical and present market growth rates suggest that the worldwide market for specialty chemicals will grow 16 percent per year (Datta, 1994).

A wide range of biobased industrial products and technologies will be introduced to diverse industrial markets. Ethanol and oxygenated chemicals derived from fermentable sugars will be key precursors to other industrial chemicals traditionally dependent on petroleum feedstocks. In the long term, with advances in genetic engineering, large-scale fuel production from lignocellulosic plant materials may become cost competitive with petroleum fuels. In other cases, biobased technologies such as enzyme catalysts are promising replacements for more hazardous industrial chemical processes. Increasingly, niche markets will be sought for a wide array of custom-engineered plant polymers (e.g., chiral compounds) not available in petrochemical-based products.

COMMODITY CHEMICALS AND FUELS

Biobased industries of the future will include plant-derived commodity chemicals (those selling for under $1.00 per pound) to provide transportation fuels and intermediate chemicals for industrial processing. Ethanol is critical because this oxygenate can serve as a transportation fuel and also is a precursor to many other industrial chemicals. For example, corn starch fermentation yields ethanol, which then can be dehydrated for production of ethylene, the largest petroleum-based commodity chemical. The U.S. Department of Agriculture (USDA) estimated for 1996 to 1997 that 12 million metric tons of corn of a total 252 million metric tons of corn grain produced in the United States were put into ethanol fuel production—about 1.1 billion gallons of ethanol fuel (ERS, 1997b).

Ethanol

Large imports of foreign crude oil in the 1960s and 1970s stimulated interest in fuel ethanol (Harsch, 1992). In the United States the primary approach taken was gasohol, a blend of 10 percent ethanol in gasoline. Researchers found that ethanol and its derivative, ethyl *tert*-butyl ether, work as octane enhancers, increasing the efficiency of burning gasoline in an internal combustion engine. Similar interest in ethanol occurred in Brazil, and, with subsidies from the government, Brazil forged ahead with ethanol production. Until six years ago nearly 95 percent of the cars produced in that country ran on ethanol. Since then the price of crude oil has dropped and Brazil has converted to ethanol-gasoline blends (Anderson, 1993).

In the United States, ethanol occupies a niche in the transportation fuel market as an oxygenate in urban areas that do not attain U.S. Environmental Protection Agency air quality standards for carbon monoxide in response to the Clean Air Act Amendments of 1990. Gasoline is blended with an oxygenate fuel such as ethanol or methyl *tert*-butyl ether (MTBE) to increase the combustion efficiency of gasoline and decrease carbon monoxide emissions in cold weather. Due to its lower cost in comparison to ethanol, MTBE has been the primary oxygenate used, and its use ranges from 63 to 81 percent of the total demand for oxygenates (EIA, 1997). Total estimated U.S. production of MTBE in 1995 was 8 billion kilograms; estimated ethanol production for 1994 was 4.3 billion kilograms (Committee on Environment and Natural Resources, 1997).

An interagency panel assessed the air quality, groundwater and drinking water quality, fuel economy and engine performance, and the potential health effects of MTBE and other oxygenates (Committee on Environment and Natural Resources, 1997). In its review of the draft federal report, the National Research Council concluded that the cold-weather air pollution effects of oxygenated fuels were unclear. While data on the occurrence of MTBE in groundwater and drinking water are scarce, MTBE has been detected in groundwater (Squillace et al., 1996), stormwater (Delzer et al., 1996), and drinking water (Committee on Environment and Natural Resources, 1997). Because MTBE is very soluble in water, is not readily sorbed by soil and aquifer materials, and generally resists degradation in groundwater, the interagency group recommended that there be an effort to obtain more complete monitoring data, behavior and fate studies, and aquatic toxicity tests for wildlife and to establish, if warranted, a federal water quality criterion.

Specific well-targeted research will be needed to answer questions about potential tradeoffs in using these chemicals as additives to gasoline (NRC, 1996b). Demand for starch-based ethanol is influenced by the commodity market price for corn. During the 1995 to 1996 marketing year, high demand for corn grain drove up corn prices to record levels, leading to high input costs and a downturn in ethanol fuel production. Many ethanol producers opted to suspend ethanol production and do maintenance on their manufacturing facilities. Other producers diverted ethanol fuel production to the alcoholic beverage market. The USDA expects that producers will need to reestablish long-term contracts with blenders to regain market share lost when corn markets experienced a period of high input pricing in 1995 to 1996 (ERS, 1997b).

In the long term, large-scale production of fuel ethanol from lignocellulose materials could become technically feasible and economically favorable. A key will be demonstrating that recent and anticipated technical innovations work at larger scales with representative raw materials. The cost of bioethanol must drop significantly if it is to penetrate a much larger fraction of the transportation fuel market. This change will occur only if economical lignocellulose conversion technologies are developed—a long-sought achievement. Use of these alternative feedstocks with new conversion processes may reduce production costs sufficiently to allow access to the commodity fuel market, even without subsidies or tax incentives. The case study of lignocellulose-ethanol processing described in this chapter illustrates one approach toward reducing the costs of ethanol production.

Biodiesel

Biodiesel is a fuel that likely will not be an economically viable product in the near term. Vegetable-based diesel fuels are appealing in part because these biobased fuels confer some potential environmental benefits. Because production costs for soy-based diesel currently are extremely high, soy-based diesel fuel faces stiff competition in most petroleum-based diesel fuel markets. For example, in Europe biobased diesel is more popular because incentives are offered to encourage its use. Further research and development may increase the demand for biobased diesel fuel in the long term.

Biodiesel is made by transesterifying plant oil(s) with methanol in the presence of a catalyst to produce fatty acid methyl esters. Methanol for the reaction is readily available from biomass, natural gas, or coal. Oils that can be processed into biodiesel include soybean, canola, and industrial rapeseed (Harsch, 1992). If the reacted oils have the correct carbon chain length, the fatty acid methyl esters will have chemical characteristics similar to those of conventional diesel fuel when they combust in modern diesel engines. Biodiesel is usually mixed with petroleum-based diesel fuel in a ratio of 20 percent biodiesel to 80 percent diesel fuel (B20). The U.S. Department of Energy (DOE) has moved to the rule-making process for inclusion of B20 as an approved alternative fuel under the Energy Policy Act of 1992. If this acceptance occurs, government-owned fleets of small diesel engines will be able to meet alternative fuel guidelines with biodiesel under that act.

Biodiesel does confer some environmental benefits. One advantage of biodiesel over petroleum-derived diesel is the virtual absence of sulfur and aromatic compounds (Abbe, 1994). Further, combustion of biodiesel produces lower emissions of carbon monoxide, unburned hydrocarbons, and particulate matter than combustion of petroleum-based diesel (Abbe, 1994). Consideration of emissions is particularly important in urban areas suffering from poor air quality. Biodiesel may be valuable in the future because the fuel can be used in today's diesel engines without modification and in various blends without negative impacts on engine performance (Hayes, 1995).

An increased focus on biodiesel largely results from its success in Europe. The crop of choice in Europe has been rapeseed, and the European Union has implemented subsidies for farmers growing oilseed crops to promote biodiesel production. European production of biodiesel and implementation of government policies to promote its use have progressed relative to the United States. A gallon of biodiesel requires 7.35 pounds of soybean oil and other inputs valued between $0.50 and $0.70. If soybean oil costs $0.25 per pound, biodiesel must cost at least $2.33 per gallon excluding taxes, or at least four times the cost of tax-free petroleum-based diesel (Hayes, 1995). The USDA estimated a hypothetical market price of $4.25 per gallon for biodiesel (ERS, 1996b). As a result of these high costs, biodiesel may be used only where it is mandated (i.e., in urban transit fleets and government-owned diesel vehicles), which limits the ultimate market size and encourages vehicle owners to seek less expensive alternatives (Hayes, 1995). Some research on other plant-based diesel fuel alternatives may be warranted. Direct substitution of plant oils for diesel fuel would be cheaper than the manufacture of biodiesel because the transesterification process imposes significant additional costs. Unfortunately, the high viscosity of the oils causes poor atomization and creates flow characteristics that are generally incompatible with present-day diesel engines (Harsch, 1992). A different lower-cost alternative that merits consideration is the use of ethanol or butanol solvents for transesterification of plant oil.

In the United States, biodiesel would be unlikely to completely replace petroleum-based diesel. Even if all of the vegetable oil currently produced in the United States (about 3.1 billion gallons per year) went into biodiesel production, plant-based diesel production could provide only 6.4 percent of the nation's annual diesel consumption of 45 billion gallons (Harsch, 1992). Production of 3 billion gallons of biodiesel necessary for agricultural uses would require farmers to dedicate 40 million to 60 million acres to biodiesel crops (Harsch, 1992). Introduction of biodiesel as a blend with conventional diesel fuel is a more feasible goal in the United States and one that could have significant benefits in areas where the environment is sensitive to disruption by conventional diesel emissions or spills.

INTERMEDIATE CHEMICALS

Intermediate chemicals play an integral role in the U.S. economy. Organic chemicals are synthesized primarily from petroleum for production of numerous nonfuel industrial products such as paints, solvents, clothing, synthetic fibers, and plastics. Without these products the United States could not maintain its modern way of life. When petroleum supplies are interrupted, price volatility occurs in international petroleum markets. These events can have widespread economic consequences on oil-importing nations. Increasing the diversity of strategic feedstocks with biobased raw materials could help mitigate economic downturns created by oil shortages. Thus, intermediate chemicals are an important market targeted by the biobased industry.

Ethylene

Ethylene is perhaps the most important petrochemical because of the value of its numerous derivatives such as polyethylene, ethylene dichloride, vinyl chloride, ethylene oxide, styrene, ethanol, vinyl acetate, and acetaldehyde. Before the new lignocellulose conversion technology came on the horizon, the ethylene market was considered inaccessible to

TABLE 3-2 Hypothetical Production Cost Comparisons for Ethylene

Commodity	Year	Average Variable Cost[a] ($/lb.)	Average Fixed Cost[b] ($/lb.)	Average Total Cost ($/lb.)	Price[c] ($/lb.)
Petroethylene	1993	0.02	0.08	0.10	0.21
Projected petroethylene	2005	0.06	0.08	0.14	
Biobased ethylene[d]	1993	0.13	0.01	0.14	

[a] Average variable costs include costs for labor, inputs, and energy in the case of biobased ethylene, but labor is omitted in the case of petroethylene because the figure was not available. This will raise the average total cost for the petroethylene somewhat but by only approximately 4 percent. In the case of petroethylene, the input material was naptha, and credit was given for the propylene and gasoline that would be coproduced.

[b] Average fixed costs include costs for land and capital.

[c] Price for ethylene on December 23, 1993, as quoted in *Chemical Marketing Reporter*. Prices will vary annually.

[d] Cost data for biobased ethylene were developed using Donaldson and Culbertson (1983) estimates of input requirements, yields, and plant costs—combining input requirements with 1993 price data to estimate material and utility expenditures, updating capital expenditure data with a price index for plant and equipment, and giving annual payment for a 15-year mortgage. The ethanol production cost of $0.46 per gallon (see Appendix A, Table A-2) converts to 6.6 pounds of ethanol per gallon = $0.12 per pound (ethanol cost for the ethylene process).

SOURCE: Gallagher and Johnson (1995).

biobased production (Lipinsky, 1981). Today, biobased ethylene production based on ethanol derived from corn stover still is not cost competitive with petroethylene sources. In the near term, ethylene based on lignocellulose fermentation could move into the margin of competition against petrochemical sources (see Table 3-2). Petroethylene costs are expected to be $0.14 per pound by 2005 based on increasing cost projections for oil prices, using long-term projections developed by the World Bank. Bioethanol costs likely will remain stable owing to a slowly growing demand for agricultural products. Ethylene would be produced in large-scale operations that already process ethanol, thus enabling manufacturers to manage the costs from sluggish marketing periods. With rising petroleum prices or further improvements in the biobased ethylene route, the cost advantage of petroethylene could erode.

Acetic Acid

Acetic acid could be a large-volume chemical target for the biobased industry. It is used primarily as a raw material in the production of vinyl acetate, acetic anhydride, cellulose acetate, acetate solvents, terephthalic acid, and various dyes and pigments and as a solvent in the chemical processing industry. The food, textiles, and pharmaceuticals industries also use acetic acid in their manufacturing processes. In 1992, 1.9 million tons of acetic acid were produced in the United States (Ahmed and Morris, 1994). Acetic acid may be combined with dolomite lime to produce calcium magnesium acetate, an important deicing agent for the transportation industry. Biobased acetic acid may be produced by fermenting corn starch or cheese whey waste or as a byproduct of the sulfite wood pulping process. A better understanding is needed of the relative costs of production of acetic acid from renewable resources as compared to petroleum-based feedstocks.

Fatty Acids

Fatty acids, readily available from plant oils, are used to make soaps, lubricants, chemical intermediates such as esters, ethoxylates, and amides. These three important classes of intermediates are used in the manufacture of surfactants, cosmetics, alkyd resins, nylon-6, plasticizers, lubricants and greases, paper, and pharmaceuticals (Ahmed and Morris, 1994). Of the approximately 2.5 million tons of fatty acids produced in 1991, about 1.0 million tons (40 percent) were derived from vegetable and natural oils; the remaining 1.5 million tons were produced from petrochemical sources. Twenty-five percent of all plant-derived fatty acids used in the coatings industry comes from tall oil (a byproduct of kraft paper manufacture). The range of compounds in tall oil is quite large and unique, including long-chain unsaturated fatty acids.

SPECIALTY CHEMICALS

Specialty chemical markets represent a wide range of high-value products. These chemicals generally sell for more than $2.00 per pound. Although the worldwide market for these chemicals is smaller than those for bulk and intermediate chemicals, the specialty chemicals market now exceed $3 billion dollars and is growing 10 to 20 percent annually (Datta, 1994). Examples of biobased specialty chemicals include bioherbicides and biopesticides; bulking and thickening agents for food and pharmaceutical products; flavors and fragrances; nutraceuticals (e.g., antioxidants, noncaloric fat replacements, cholesterol-lowering agents, and salt replacements); chiral chemicals; pharmaceuticals (e.g., Taxol); plant growth promoters; essential amino acids; vitamins; industrial biopolymers such as xanthan gum; and enzymes.

Specialty chemicals can be made using fermentation and enzymatic processes or directly extracted from plants. Genetic engineering has now made possible microbial fermentations that can convert glucose into many products and can yield an essentially unlimited diversity of new biochemicals (Zeikus, 1990). Likewise, one could engineer plants to produce some of these same chemicals. Furthermore, industrial researchers are discovering that plants can be altered to produce molecules with functionalities and properties not present in existing compounds (e.g., chiral chemicals). It is anticipated that advances in biotechnologies will have significant impacts on the growth of the specialty chemicals market.

Enzymes

Fermentation of biological materials will continue to be a primary source of most enzymes used today and new enzymes produced in the future. Enzymes serve two major purposes. Some function as biological catalysts in industrial processing of food ingredients, specialty chemicals, and feed additives. Others are components in end products such as laundry detergents, diagnostics, laboratory reagents, or digestive aids.

Worldwide enzyme sales totaled $650 million in 1989 (Layman, 1990) and grew to approximately $1 billion in 1993 (Thayer, 1994). European companies dominate world enzyme production; the largest company, Novo Nordisk, currently supplies 40 to 50 percent of world sales (Thayer, 1994). Analysts predict enzyme sales will grow 10 percent annually over the next few years for traditional markets and new uses. The three largest markets for enzymes are the detergent, starch, and dairy industries. The enzyme market in 1989 broke down into 40 percent for detergents, 25 percent for starch conversion, and 15 percent for dairy applications (Layman, 1990). The remaining 20 percent included leather, pulp and paper, and animal feed manufacture. This last category is of particular interest because it includes industries that historically have caused adverse environmental impacts and, consequently, may have incentive to use more environmentally benign processes like those based on enzymes.

Soaps and Detergents

Industrial production of soaps and detergents in the United States totaled $14.9 billion in 1993 (Ainsworth, 1994). Almost half of the laundry detergents in the United States and 90 percent of those in Europe and Japan contain enzymes. The partial ban in the United States of water-polluting phosphates from detergents in 1982 led to increased use of enzymes in soaps and detergents (Ahmed, 1993). The replacement of traditional chlorine bleach with peroxygen-based bleach additives (such as perborate bleach) also has enabled enzymes to play an important role in the soap and detergent industry due to their compatibility with the newer additives (Ainsworth, 1994).

Enzymes are naturally diverse and function in various cleaning agent roles. Protease, lipase, and cellulase enzymes are used in soaps and detergents to break down and help remove dirt stains. Celluzyme, a Novo Nordisk product, removes microfibrils that emerge from cotton fibers after use and cause an "old and gray" appearance (Falch, 1991). In addition, detergent enzymes reduce energy use because they are effective in much cooler wash waters.

Food Processing Enzymes

The largest use of enzymes as catalysts is in the production of high-fructose syrup from starch. Amylases break down starch to glucose; then glucose isomerase is used to isomerize the glucose into fructose. The resulting mixture of glucose and fructose is used as a sweetener in soft drinks. Enzymes also have several uses in the dairy industry. The enzyme rennin coagulates milk protein and is used to make dairy products such as cheeses. Lactase is used to produce lactose-free milk.

Cellulase Enzymes

Relatively small amounts of cellulase enzymes are used now, primarily in the food industry. A large-scale fermentation industry based on lignocellulosic materials will require huge volumes of cellulases, much larger amounts than for any other enzyme, at much lower enzyme prices than currently available. Reducing the costs of cellulase enzymes is a key research priority for reducing the costs of industrial processing of biobased raw materials.

Other Uses for Enzymes

Various industries use enzymes as end products or biocatalysts at a smaller scale. The leather manufacturing industry has traditionally used lime and sodium sulfide mixtures to dissolve hair on animal skins—a process that is polluting and unpleasant to work around. Proteases provide an alternative treatment that loosens and removes the hair, allowing it to then be filtered off. Proteases also result in a better-quality end product (Falch, 1991). The pulp and paper industry also uses enzyme technologies, especially xylanases for bleaching to replace chlorine. The textile industry uses cellulases for making "stonewashed" jeans (Wrotonowski, 1997).

Animal Feed Industry

The animal feed industry is currently developing beneficial applications of enzymes on a large scale. Certain antinutritional compounds are present in animal feeds. Beta-glucans create viscous mixtures after being solubilized, and these impair animal digestion by causing poor absorption, poor diffusional rates of solutes in the digestive tract, and a low rate of nutrient uptake. Addition of beta-glucanases (enzymes that degrade beta-glucans) to animal feeds removes the beta-glucans and their associated problems. This technique makes it possible to produce efficient animal feeds from grains that are high in beta-glucans, such as barley and oats. It also decreases the amount of manure produced by animals consuming the feed.

Enzymes may lessen the contribution of animal feeds to phosphate pollution. Phytic acid, the major plant storage compound for phosphate, comprises about 60 to 65 percent of the phosphorus content in animal feeds made from cereal grains. Phytic acid forms complexes with iron and zinc ions and makes these metal ions less available for assimilation by animals. Moreover, animals cannot degrade phytic acid, so producers add inorganic phosphate to animal feed as a supplement, although most of the supplemental phosphate is excreted. The estimated 100 million tons of animal manure produced each year in the United States is thought to liberate 1 million tons of phosphates, contributing significantly to phosphate pollution. Addition to animal feeds of phytase, an enzyme that degrades phytic acid, allows animals to digest the phytic acid and better assimilate the iron and zinc ions. Less phosphorus consequently needs to be added to the feed, thereby reducing the contribution of animal feed to phosphate pollution. The animal feed industry can benefit from the addition of a variety of enzymes to feed mixes. It is important to note that modifications of feed composition can be made through genetic engineering of plants to allow for optimization of the feed directly.

BIOBASED MATERIALS

Increased consumer demands for environmentally benign products are leading to numerous opportunities in the biobased materials market. Diverse materials are produced from agricultural feedstocks, including wood and paper; cotton, kenaf, and other textiles; industrial starches; and specialty polysaccharides such as xanthan, fats and oils, and proteins (Narayan, 1994). Also under development are biobased composites such as one made of soybean protein and waste paper.

Bioplastics

Renewable resources such as industrial starches, fatty acids, and vegetable oils can serve as sources for bioplastics. Biodegradable thermoplastics—such as starch esters, cellulose acetate blends, polylactide, thermoplastic proteins (e.g., zein), and poly(hydroxybutyric acid)

BOX 3-1
Plastics from Plants and Microbes

Poly(hydroxybutyrate) (PHB) and its variants, generally known as polyhydroxyalkanoates, are natural polymers commonly produced by plants and microbes. PHB is a truly biodegradable plastic material that is naturally and efficiently degraded to carbon dioxide and water by many common soil bacteria. It is also a common food storage material in bacteria that accumulates inside bacterial cells when carbon is in excess but some other nutrient limits growth. In bacteria such as *Alcaligenes eutrophus*, PHB may represent as much as 90 percent of the total cell mass under the appropriate growth conditions. PHB is derived from acetyl-CoA, a component of primary metabolism, by a process involving three enzymes. The enzyme beta-ketothiolase condenses two molecules of acytl-CoA to yield acetoacetyl-CoA. This compound is reduced to beta-hydroxybutyryl-CoA by acetoacetyl-CoA reductase and then condensed to a nascent polymer chain by PHB polymerase.

Commercial production by fermentation is currently under way for poly-3-hydroxybutyrate-3-hydroxyvaleate (PHB-V), a PHB that has characteristics similar to polypropaline or polyethelene. Nevertheless, a broader commercial use of these natural polymers will require new biological and engineering technologies to enable large-scale production. One possible approach is to increase and improve synthesis in the host bacterium by mutating the existing biosynthetic pathway. Another is to move the genes for PHB synthesis into other bacteria, plants, or yeasts for increased production. At Michigan State University, Sommerville and colleagues pioneered this approach using the common weed *Arabidopsis* as a "bioreactor." The researchers manipulated genes for PHB synthesis in *Arabidopsis* and showed the plant produced PHB at a low level. Moving the genes to the target expression in a different part of the plant cell (i.e., from the cytoplasm to the chloroplast) dramatically increased PHB production.

Increased understanding of the basic science underlying the plant and bacterial metabolic and biosynthetic pathways has made possible another exciting development. New polymer structures can now be engineered by manipulating the PHB metabolic pathway in various plants and microbes. For example, less brittle plastics may result if low amounts of poly(hydroxyvalerate) are coproduced in bacterial or plant cells that manufacture PHB. These polymers are readily biodegradable and have properties that make them a suitable substitute for petrochemical-derived thermoplastics. The new-found ability to "engineer" chemical derivatives of such polymers in living cells holds the promise of a truly environmentally benign bioprocess.

SOURCE: Poirier et al. (1995).

(PHB)—show great promise for replacing the plastics derived from petrochemicals that generally are not biodegradable (see Box 3-1). Graft plastic polymers (plastics based on plant materials and petrochemicals) are less bio- degradable than plant-based bioplastics. Bioplastics comprise about 5 percent of the total polymer, plastics, and resin market (Ahmed and Morris, 1994).

The bioplastics industry has generated new markets for industrial starches. Starch can be directly manufactured into products such as biodegradable loose-fill packaging to replace nondegradable polystyrene-based packaging peanuts. Fermenting starch into lactic acid or PHB yields other starch-derived thermoplastics. The Cargill Company has introduced polylactide-based thermoplastics for single-use disposable products such as utensils, plates, and cups. ICI Corporation has commercialized biodegradable PHB plastics for shampoo bottles and other higher-cost disposables. Plant matter also provides a new material for direct processing into plastic and polymeric resins.

A graft copolymer of latex and starch is used to make coated papers. Certain starch-based plastics are also in commercial use, as are various graft polymers between starch and synthetics. One class of graft polymers absorb many times its weight in water and has many applications such as absorbent soft goods (e.g., absorbents for body fluids, disposable diapers, hospital underpads, and related products), hydrogels, and agricultural products (such as seed and bare root coatings and hydromulcher) (Doane et al., 1992). These hydrophilic graft polymers are prepared using polyacrylonitrile in which the nitrile substituents have been hydrolyzed with alkali. Many new starch-based polymers and applications are expected to appear soon in commercial uses.

Soy-based Inks

Soybean oil-based inks were introduced to U.S. markets in the 1970s in response to the oil shortages. More recently increased emphasis on improving worker safety and reducing environmental emissions has spurred interest in alternatives to petroleum-based inks. Soybean oil is a carrier for a pigment in ink formulations. Plant-derived inks require less use of hazardous chemicals during equipment maintenance, produce lower evaporative emissions of volatile organic hydrocarbons, and are biodegradable. Soybased inks are more desirable because the lighter color of soybean oil enhances the true color of colored ink pigments compared to petroleum-based inks. Black soy-based inks typically require a larger proportion of oil than pigment in comparison to colored printing inks. Because soybean oil costs more than petroleum, black soybased inks are at a cost disadvantage. Some research indicates that soy inks can spread about 15 percent farther than petroleum-based inks, offsetting the differences in cost (EPA, 1994).

Forest Products

Forest products currently comprise the largest source of renewable resources for the biobased materials industry. The United States annually produces more than 260 million tons of lumber, paper, and derived wood products with a combined value exceeding $131 billion (Chum and Power, 1992). Currently, wood and paper products account for $96 billion or 87 percent of agricultural and forest products used in industrial production (ERS, 1997b). Solid wood is used for lumber and many other products and is the major building material in the United States. Approximately half of the 1993 U.S. wood harvest, 266 of the 501 million cubic meters harvested, was initially sent to solid wood products mills. The mill residue remaining was sent to pulp mills or used for fuel. Overall, approximately 28 percent of the 1993 harvest ended up in solid wood products (Ince, 1996). Preexisting uses of silviculture crops in the pulp and paper industry may provide predictable markets and sustain production of woody silviculture crops while new uses develop in the manufacture of high-value materials and chemicals.

The pulp and paper industry is one of the largest industries in the United States, producing 64 million tons of pulp and 88 million tons of paper in 1991 (imports, exports, inorganic fillers, and recycled pulp account for the difference in tonnage between pulp and paper; Miller Freeman, Inc., 1995). Recent innovations may expand the applications of cellulose pulp. The U.S. Forest Products Laboratory developed a "spaceboard," a lightweight structural composite made by molding pulp fiber slurries into waffle-shaped forms (Hunt and Scott, 1988) that is now commercialized. Scientists have also developed moldable plastic materials by combining pulp fibers with thermoplastics. These materials have many potential uses, for example, in car bodies and packaging materials.

Courtaulds, a company from the United Kingdom, and Lenzing, an Austrian firm, each have begun large-scale production of lyocell, a cellulosic fiber made from a solvent spinning process and sold under the trade name of Tencel. Like rayon, Tencel is wood-pulp based; however, rayon requires dry cleaning and Tencel is washable. Tencel is the first new textile fiber to be introduced in 30 years and has been described as the "best thing since cotton."

Cotton and Other Natural Fibers

Cotton is one of the most promising industrial crops. Cotton is a

RANGE OF BIOBASED PRODUCTS   69

> **BOX 3-2**
> **Biopolymers**
>
> The great majority of all biomass consists of natural polymers, and the great majority of all biomass is carbohydrate in nature. This means that the majority of all biomass is in the form of carbohydrate polymers, called polysaccharides. These natural polymers (biopolymers) can be used both as nature provides them and as the skeletal framework of other derived polymers. By far the most abundant of these carbohydrate polymers is cellulose, the principal component of cell walls of all higher plants. It is estimated that 75 billion tons of cellulose are biosynthesized and disappear each year, most of the disappearance being through natural decay.
>
> Cellulosic plant materials are used as fuel, lumber, and textiles. Cellulose is currently used to make paper, cellophane, photographic film, membranes, explosives, textile fibers, water-soluble gums, and organic-solvent-soluble polymers used in lacquers and varnishes.
>
> The principal cellulose derivative is cellulose acetate, which is used to make photographic film, acetate rayon, various thermoplastic products, and lacquers. The world's annual consumption of cellulose acetate is about 750,000 tons, 400,000 to 450,000 tons being produced in North America. Cellulose acetate products are biodegradable.
>
> While use of biopolymers, largely polysaccharides, as is and in modified form is now considerable, only a infinitesimal amount of that available is now utilized commercially in applications also served by petroleum-based polymers; so the potential is enormous. Broader application of such preformed polymeric materials awaits research and development.

plant fiber composed of 90 percent cellulose. The long fibers of cotton make it an ideal material for weaving and spinning into cloth. Cotton confers qualities on fabrics that are difficult to duplicate with synthetic fabrics. Demand for cotton products has resurged in recent years, and the United States harvested approximately 18 million bales of raw cotton in 1996 to 1997 marketing year (USDA, 1997a). Advances in biotechnology and genetic engineering are now enabling development of cotton cultivars with improved pest resistance, yield, and quality, thereby potentially reducing production costs and better matching cotton characteristics to specific applications.

Natural fibers other than cotton occupy various U.S. niche markets, such as specialty fabrics, papers, cordage, and horticultural mulches and mixes. Heightened environmental concerns are helping natural fibers find their way into new markets as well. Jute, hemp, sisal, abaca, coir fibers, and products derived from these fibers are currently being imported but could be produced domestically.

TARGETING MARKETS

This section identifies opportunities for replacing products made from nonrenewable fossil feedstocks with biobased chemicals and materials. In the midterm, biobased products will be primarily oxygenated chemicals and materials; petroleum will remain a more competitive feedstock for hydrocarbon-based liquid fuels and aromatic and alkane chemicals. Over the longer term, adoption of biofuels and biobased aromatic and alkane chemicals could grow significantly, given investment in the necessary research and development and perhaps carefully chosen incentives. The manufacture of chemicals is now dominated by fossil fuel sources; this market may represent the greatest opportunity for replacement of petrochemicals with biobased material. Only about 10 percent of the 100 million metric tons of chemicals marketed in the United States are biobased. The remaining 90 million tons of organic chemicals currently derived from fossil fuels potentially could be replaced by renewable resources.

Biomass processing—fermentation of starch or cellulose accompanied by additional chemical, thermal, and physical processing steps—can produce a number of oxygenated intermediate chemicals, including ethylene glycol, adipic acid, ethanol, acetic acid, isopropanol, acetone, butanol, citric acid, 1,4-butanediol, methyl ethyl ketone, N-butanol, succinic acid, itaconic acid, lactic acid, fumaric acid, and propionic acid. These intermediate chemicals have uses in the manufacture of such polymers as nylon, polyesters, and urethanes; of various plastics and high-strength composites; and of solvents, coatings, and antifreeze. Numerous chemical markets may be filled by agricultural feedstocks.

Development of fermentation industries over the next few decades could be accomplished in three progressive phases: (1) glucose from corn feedstock, (2) sugars from lignocellulosic crop residues from silviculture and agriculture, and (3) establishment of a large "carbo-chemistry" industry using sugars derived from the most cost-effective local sources. Another approach would be to produce industrial chemicals in genetically engineered plants. These chemicals and materials could then be separated from the plant matter and upgraded by processing, possibly biological processing in some cases.

Apart from pulp and paper, ethanol fuel is probably the largest single biobased industrial product. The United States produced about 1.1 billion gallons of ethanol in the 1995 to 1996 marketing year (ERS, 1997b), less than 1 percent of annual domestic gasoline consumption. The United States used 1.8 billion tons of mostly fossil fuels in 1989, triple the amount of all of the plant matter consumed for food and nonfood purposes combined (Ahmed, 1993). Large-scale production of ethanol fuel from lignocellulose may become economically feasible with new processing technologies, although this may take decades to materialize.

Specialty chemicals represent a rapidly growing and diverse group of high-value industrial products. The benefits of some of these biobased products are well known (e.g., enzymes). At the same time, rapid advances occurring in the life and materials sciences will lead to discoveries of plant compounds that cannot be produced with petroleum feedstocks. Industry will vigorously pursue the most promising candidates for further development and commercialization. Since some of these products will be successful and others will not, this market will be constantly evolving. It will be important for academic and industrial scientists to monitor market trends and technological breakthroughs to identify promising target areas for future research.

Significant opportunities exist to increase markets for biobased materials. Because most industrial materials can be produced from agricultural and forestry feedstocks, markets for biomaterials and biopolymers likely will increase. For example, the United States manufactures annually more than 5 billion pounds of industrial starches for multiple uses, including making paper and paperboard. Several biodegradable polymers—such thermoplastics as polylactide and poly(hydroxybutyrate)—have been developed and are now sold commercially in small quantities. It is likely that biobased materials, including biopolymers, will serve as important and diverse resources for a growing biobased industry.

CAPITAL INVESTMENTS

A substantial investment of capital will be required to commercialize biobased products. Capital investment figures estimated for nine biobased chemicals are shown in Table 3-3. The private sector currently is investing in lactic acid production for use in lactic acid polyester formation, a polymer that can substitute for polystyrene in many cases. The capital required for developing the remaining eight chemicals would be more than $6 billion.

Scale-up and commercialization costs are significant barriers in moving laboratory discoveries into the market. Industrial researchers estimate that the relative costs of discovery, scale-up, and commercialization are 1:10:100. Hence, $1 million invested in basic research generates sufficient promising technologies to justify $10 million invested in scale-up and risk reduction efforts that, in turn, are sifted through to find sufficient proofs of concept to warrant $100 million invested in commercial-scale manufacturing facilities. Applying these ratios to the eight chemicals listed in Table 3-3 suggests that $600 million dollars would be required to adequately demonstrate promising production technologies at a pilot scale and that $60 million would be required to support research aimed at solving the technical issues that have the greatest impact on processing costs.

TABLE 3-3   Estimated Capital Requirements for Target Biobased Organic Chemicals Produced from Glucose

Biobased Organic Chemical	1993 Output (Million pounds)	Capital Requirements ($ millions)[a]	Corn Required (Bushels per day)
Acetic acid	3,658	1,350	378,000
Acetone	2,462	1,221	342,000
Butanol	1,328	1,157	342,000
Maleic anhydride	424	230	60,000
Methyl ethyl ketone	556	484	126,000
Isopropanol	1,236	1,084	303,000
Butanediol	200	196	42,000
Adipic acid	760	230	60,000
Lactic acid	10,063	2,208	882,000

[a] Capital requirements and development and commercialization costs were updated to reflect 1993 prices. Model uses estimated costs for input requirements, yields, and plant costs—combining input requirements with 1993 price data to estimate material and utility expenditures, updating capital expenditure data with a price index for plant and equipment, and giving annual payment for a 15-year mortgage.

SOURCE: Gallagher and Johnson (1995), based on model developed by Donaldson and Culberson (1983).

A CASE STUDY OF LIGNOCELLULOSE-ETHANOL PROCESSING

Corn stover—the stalks, leaves, and husks of corn—is a suitable feedstock for the process of converting lignocellulosics to ethanol and may provide a particularly low-cost input in the Midwest. A 1993 study by the DOE concluded that ethanol could have a price comparable to the wholesale price of gasoline if it were processed from wood chips at a large plant (Bozell and Landucci, 1993). This case study adapts the DOE cost study of wood chips to a process based on corn stover and examines how the supply of corn stover, process yields, and material flow can affect processing costs. The analysis assumes that processes for conversion of the lignocellulosics of corn stover to ethanol will be developed successfully, considers the potential impacts of transportation costs on the location of a large plant in the Midwest, and assumes that sufficient residues are left on the field to meet soil conservation goals. Eventually, a significant share of the U.S. fuel supply—7 percent of U.S. liquid fuel consumption—could be provided by ethanol produced from corn stover. The economic analysis in Appendix A suggests that up to 7.5 billion gallons of ethanol could be produced annually at a cost of about $0.46 per gallon. When corrected for fuel efficiency, the cost to replace a gallon of gasoline becomes roughly $0.58, potentially making the cost of ethanol competitive without subsidies. An additional 4.5 billion gallons of ethanol may be produced at higher costs due to potentially higher prices for corn stover when the corn stover for alcohol has to compete with its use for animal feed. Several developments will need to occur before stover-based ethanol production becomes practical. Putting these findings into practice will require scale-up and demonstration of the technology at the pilot scale to confirm rates, yields, and performance on real feedstocks for extended periods. Additionally, the real cost of corn stover must be verified through actual collection and utilization tests. Additional research would be necessary to determine the competitiveness among various feedstocks, such as residues, plant byproducts, and crops from marginal low-rent areas. The implications of widespread harvesting of crop residues on soil and water quality is another area for investigation. Finally, upstream handling, pretreatment, and lignocellulose conversion technologies must be demonstrated before investment and operating costs can be more precisely established. Large-scale production of biobased ethanol may be a long-term possibility, but some major technical barriers still need to be overcome to reduce costs.

Until now technologies for converting lignocellulosics to ethanol via hydrolysis and fermentation have not aroused commercial interest. However, recent advances may make practical the simultaneous conversion of cellulose and hemicellulose to ethanol at comparable fermentation rates, thereby enabling essentially complete conversion of lignocellulosic carbohydrates to ethanol (Ingram et al., 1987; Zhang et al., 1995). The new technology incorporates simultaneous saccharification and a recombinant microorganism for fermentation of 5- and 6-carbon sugars. The demonstration phase is just starting and may require as long as five years to confirm or refute current high expectations.

A key to the production of fuel ethanol from lignocellose will be to demonstrate that recent technical innovations work at larger scales with representative raw materials. Apart from pulp and paper, ethanol fuel is probably the largest single biobased industrial product. In recent years the United States has generated over 1 billion gallons of fuel ethanol annually from corn starch, less than 1 percent of annual domestic gasoline consumption. Today's cost of bioethanol must drop significantly if it is to penetrate a much larger fraction of the transportation fuel market. This change will occur only if economical lignocellulose conversion technologies are developed—a long-sought achievement but one that is also much nearer than it was two decades ago.

4

Processing Technologies

Expansion of biobased industrial production in the United States will require an overall scale-up of manufacturing capabilities, diversification of processing technologies, and reduction of costs. The development of efficient "biorefineries" that integrate production of numerous biobased products would help reduce costs and allow biobased products to compete more effectively with petroleum-based products. The development of new or improved low-cost processing technologies will largely determine which biobased products become available. Currently, certain processing technologies are well established while others show promise but will require additional refinement or research before they come into practical use. U.S. production of low-cost agricultural materials and experience with prototype biorefineries position the nation well to capitalize on such processing improvements.

The market prices of large-scale (commodity) biobased industrial products will depend on two primary factors: (1) the cost of the biobased raw material from which a product is made and (2) the cost of processing technology to convert the raw material into the desired biobased product. The industries for producing chemicals and fuels from petroleum are characterized by high raw material costs relative to processing costs, while in the analogous biobased industries processing costs dominate. Reducing the costs of producing commodity biobased industrial products such as chemical intermediates derived from fermentation will depend most strongly on reducing the costs of processing technology, the focus of this chapter.

THE BIOREFINERY CONCEPT

Today's petroleum refineries generate numerous products efficiently and at a very large scale from crude oil, an inexpensive raw material. The development of comparable biorefineries will be essential to make many biobased products competitive with their fossil-based equivalents. Each biorefinery would process essentially all of its biobased feedstock into multiple value-added products. The product types would include not only those manufactured by petroleum refineries but also many others that petroleum refineries cannot produce. Some examples of known and potential biorefinery products are:

- fermentation feedstocks (starch, dextrose, sucrose, cellulose, hemicellulose, molasses);
- food products (oil, starch, sweeteners);
- nonfood industrial products (loose fill packing material, paper sizes, textiles sizes, adhesives);
- chemical intermediates (lactic, acetic, citric, and succinic acids);
- fuels (ethanol, acetone, butanol);
- solvents (ethanol, acetone, butanol, esters);
- industrial enzymes; and
- biodegradable plastic resins.

Existing U.S. Prototypes

Prototypes of highly integrated biorefineries already exist in the United States. Plants that currently process agricultural and forestry materials into value-added products include corn wet and dry milling plants, soybean processing plants, wheat mills, and paper mills. The wheat, soybean, and corn operations are highly efficient and process over 95 percent of incoming feedstocks into value-added products. In some places, industrial complexes centered around corn wet milling use a single feedstock, corn, to produce a variety of products. Similar refinery complexes could be developed around corn dry mills or fermentation ethanol plants. Today's paper mills, wood products plants, and sugar beet refineries are partially integrated systems. Several pulp and paper mills produce pulp, paper, lignin byproducts, and ethanol while recycling waste paper—all on a single site with leftovers being dewatered and burned to produce steam and electricity. They have the potential to become more fully integrated by further processing, thereby enhancing the value to consumers of all their outputs. Waste paper and municipal sludge are examples of feedstocks around which biorefineries might one day be built, although this development has not yet occurred.

Corn Wet Milling

Corn wet mills used 11 percent of the U.S. corn harvest in 1992 (worth $2.6 billion on the grain market), made $7.0 billion of products, and employed almost 10,000 people (Agricultural Census 1994). Wet milling of corn yields corn oil, corn fiber, and corn starch (Figure 4-1). The starch products of the U.S. corn wet milling industry are fuel alcohol (31 percent), high-fructose corn syrup (36 percent), starch (16 percent), and dextrose (17 percent). Although a greater volume of the starch goes to non-food uses, food uses (excluding syrups) are more significant in terms of dollars per pound. Corn wet milling also generates other products (sometimes called "coproducts")—gluten feed, gluten meal, oil, and steepwater (Hacking, 1986).

Most corn fed into a wet mill goes to the primary products (starch and oil), and the balance goes to other products; the relative proportion depends on the corn's initial moisture content (Figure 4-1). High-quality No. 2 yellow dent corn currently is the preferred feedstock for wet mills

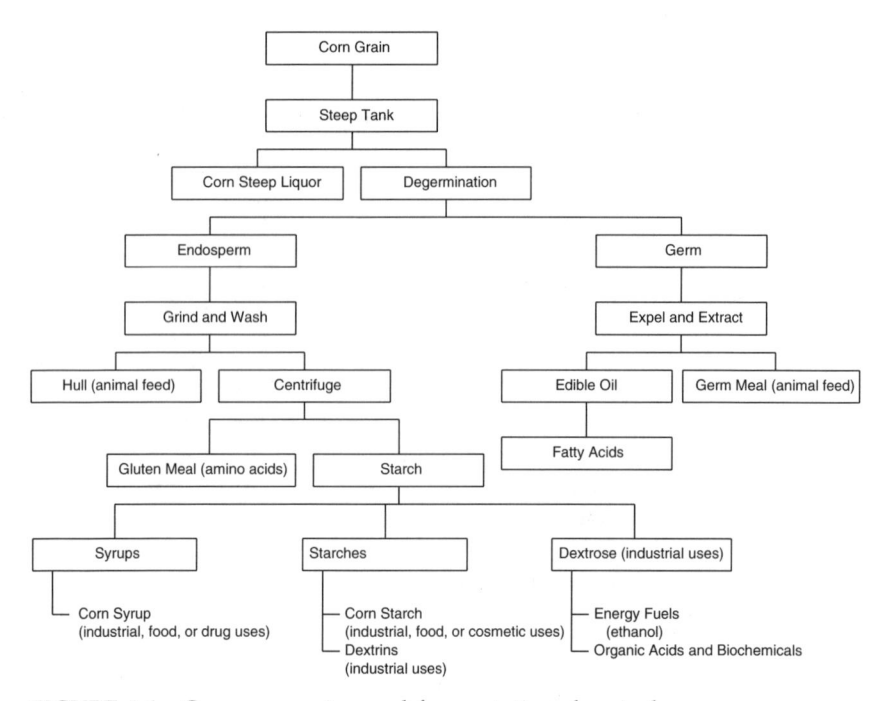

FIGURE 4-1 Corn processing and fermentation chemicals.

because it results in the highest yield of starch and oil relative to the lower-value coproducts. North American corn gives the best yields in comparison to European, African, and Asian corn varieties. Latin America could produce corn of equal quality to North American corn.

Wastes generated by corn wet milling are relatively benign and readily treated onsite. However, these wastes also represent potential fermentation feedstocks for generating additional value-added products.

Corn wet mills could comprise the front end of an industrial complex that produces food, specialty chemicals, industrial products, fuels, and pharmaceuticals. Such an expanded biorefinery would provide cleaner and more economical processes for producing existing products, new intermediates for manufacturing new products, an expanded stable market for wet millers, and an expanded market for corn farmers.

The Archer Daniels Midland (ADM) complex in Decatur, Illinois, is the prototype for such an expanded biorefinery. There, a large corn wet-milling plant and a steam and electric cogeneration station form the nucleus for several other plants. The wet mill is the source of materials for plants that produce industrial enzymes, lactic and citric acids, amino acids, and ethanol. The enzymes are then used to convert starch to lower-molecular-weight products, principally various maltodextrins and syrups (i.e., in liquefaction). The lactic and citric acids are used in processed foods, detergents, and polymers. The amino acids are used as feed and food supplements and, in the case of phenylalanine, to make aspartame. The ethanol is used as a fuel or an industrial solvent.

The United States is well positioned to develop biobased industries according to the above model of a biorefinery complex having corn wet milling as its nucleus. The current U.S. corn wet-milling industry and U.S. production of preferred corn feedstocks could provide the springboard. In addition, certain corn hybrids under development may yield higher levels of starch. This could lower feedstock costs and decrease byproduct production if these hybrids were more widely planted and used for industrial purposes. In the 1996 to 1997 marketing year, about 7 percent of U.S. corn grain was processed into industrial products (e.g., fuel alcohol; refer to Table 4-1); the remainder was used for food and animal feed.

Soybean Processing

The United States is the largest grower and processor of soybeans in the world. Soy processing yields 80 percent products (i.e., soybean oil, soy protein, lecithin, and soy protein hydrolyzate) and 20 percent defatted soy meal coproduct (see Figure 4.2; Szarant, 1987). Most of this defatted meal

TABLE 4-1 Industrial and Food Uses of Corn, 1996 to 1997 Marketing Year[a]

Product	Application	Total Food and Industrial Uses (million bushels)
High-fructose corn syrup	Food	505
Glucose and dextrose	Food	240
Cereals	Food	120
Starch	Food	39
	Industrial	186
Alcohol	Beverage	70
	Manufacturing	60
	Fuel	435
Total industrial use (excludes food and animal feed uses)		681
Total industrial and food uses		1,655
Total corn grain produced		9,265[b]

[a] Industrial uses of corn grain as raw material in the manufacturing of industrial products include industrial starch, manufacturing alcohol, and fuel alcohol. Food uses include grain processed for edible applications including high-fructose corn syrup, glucose and dextrose, cereals, food starch, and beverage alcohol.

[b] Datum is the total grain produced for food, animal feed, and industrial uses for the 1996 to 1997 marketing year.

SOURCE: ERS (1996b); USDA (1997a).

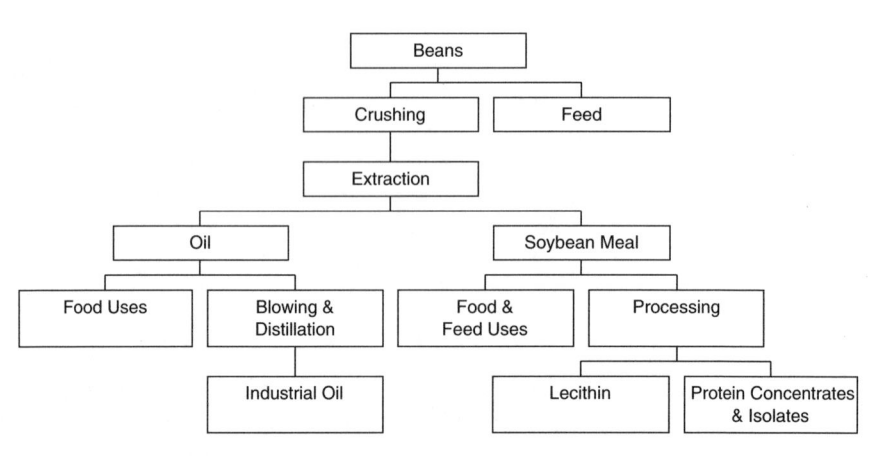

FIGURE 4-2 Soybean processing.

ends up as animal feed. The U.S. industry is internationally competitive and stable. Moreover, all of the wastes from soy processing plants are relatively benign biological materials that can be readily degraded in conventional waste treatment plants and are usually processed onsite.

As with corn wet milling, soy processing plants could serve as the front end of an industrial complex that produces food, specialty chemicals, fuels, and pharmaceuticals. North American soybean growers are highly efficient and outperform international competitors, except where the competition is supported by local tariffs. American and Canadian soybean production and processing technologies are modern and very advanced. As with corn, such characteristics position the United States well for developing industrial biorefineries centered on soy processing. Such expansion would, however, require new uses for soy protein, since existing protein markets in animal feeds are saturated and could not absorb additional production without depressing prices. In the future, genetic engineering techniques may be used to alter the soy proteins, leading to expanded uses and increased value of industrial processed soybeans.

Comparison of Biorefineries to Petroleum Refineries

The comparison of biorefineries to petroleum refineries in Table 4-2 suggests that biorefineries offer a number of potential advantages because they rely on sustainable, domestically produced raw materials.

The development and expansion of biorefineries could yield great benefits to the public at large (see also Chapter 1). Such biorefineries could produce significant amounts of valuable products from domestic renewable resources and consequently reduce national dependence on fossil feedstocks. This change could also reduce the level of pollutants generated by industrial production while still providing goods at prices and qualities comparable to those derived from nonrenewable resources. Further research is necessary to examine environmental and energy impacts at all phases of manufacturing. Switching industries to the efficient use of renewable feedstocks through biorefineries will take time and require considerable technical and financial effort. The quickest pathway would be to build on the existing corn wet mills. Cargill and ADM already have several corn wet-milling facilities that approach being integrated biorefineries. Lessons learned from these operations and from refining fossil feedstocks could help in upgrading other large grain mills and soybean processors to biorefinery status. With additional effort, paper, sugar, and wood products manufacturers could be brought online. In the future, biorefineries could use other feedstocks such as municipal sludge and mixed waste paper, crop residues, or dedicated lignocellulosic crops such as poplar or switchgrass.

TABLE 4-2 Comparison of Biorefineries to Petroleum Refineries

Aspect of Comparison	Biorefineries	Oil Refineries
Impact on primary producers	Benefits U.S. farmers	Benefits U.S. and foreign producers
Impact on primary processors	Benefits U.S. processors	Benefits U.S. and foreign refineries
Impact on other users	Gives fuels, food, pharmaceuticals, specialty and commodity chemicals producers more options at potentially lower costs	Status quo
Technical stage	Early, room for tremendous improvement	Mature, not much room for improvement
National security	Less dependence on foreign feedstocks	Greater dependence on foreign feedstocks
Export potential	Potential to export more finished goods from domestic resources	Increases import of primary and finished goods
Environmental effects	Largely positive to neutral	Many negatives

Lessons from Petroleum Refinery Experience

Several lessons from fossil feedstock refineries might prove helpful in the future development of biorefineries and should be incorporated into strategic planning for the industry. These lessons include the following:

- refineries invariably produce more and more products from the same feedstock over time, thereby diversifying outputs;
- refineries are flexible and can shift outputs in response to change
- processes in refineries improve incrementally over time; and
- process improvement invariably makes the cost of raw material the dominant factor in overall system economics.

Developers and analysts of biorefineries can use the above criteria to measure progress toward fully developed biorefinery systems. The economic and technical performance of relatively new biorefinery systems will virtually always improve over time.

PROCESSES FOR CONVERTING RAW MATERIALS TO BIOBASED PRODUCTS

Biorefineries of the future will use technologies based on thermal, mechanical, chemical, and biological processes to derive industrial products from renewable resources. Conversion of raw materials—such as wood, other lignocellulosic biomass, vegetable crops, starch-producing crops, and oil seeds—to final end products often will require a combination of processes.

This section identifies the technologies that might be used by future biobased industries. A distinction is made here between "proven" and "potential" technologies. The former covers technologies that have been evaluated at a large enough scale to ensure the process is technically practicable in a commercial plant. "Potential" technologies, in contrast, are generally those that have been evaluated at laboratory scale and are sufficiently attractive to be of potential commercial interest.

Lignocellulose Fractionation Pretreatment: A Key Step

Biological conversion of carbohydrates by fermentation and enzymatic processes is perhaps the most flexible method of converting renewable resources into industrial products. The carbohydrates for most fermentation-derived products currently come from corn starch. Without new carbohydrate sources, the cost and availability of starch ultimately would determine the scale of biobased industries based on carbohydrates. Lignocellulosic materials could potentially provide a new, much larger (by at least two orders of magnitude), and less expensive carbohydrate source for biobased industries. Realization of this potential will depend on the development of inexpensive and effective processing technology to fractionate and convert lignocellulosics to fermentable sugars (Lynd, 1996). Despite past investigations of many processes, none has yet enabled growth of a large-scale bioconversion industry based on lignocellulosics (McMillan, 1994). The necessary technical advances present a formidable problem but one that is appropriately a high research and development priority for the nascent biobased products industry.

Utilization of all three fractions of the lignocellulose—the lignin (15 to 30 percent), the cellulose (35 to 50 percent), and the hemicellulose (20 to 40 percent)—is desirable from a "no waste" engineering design maxim, just as utilization of the oil, protein, and fiber of corn and the steepwater of the wet-milling process is essential in a starch-based industry. Thus, processes discussed below include those for lignocellulose fractionation and those for processing the component fractions.

Proven

Removing the lignin from lignocellulose is one possible step toward obtaining carbohydrate for further bioconversion. Lignin makes up about 15 to 30 percent by weight of lignocellulosics such as wood and the woody parts of annual plants. Because paper production from this raw material requires removal of most lignin, the pulp and paper industries have well-developed processes for this purpose. Chemical wood pulping produces as a byproduct more than 50 million metric tons of lignin a year in the United States. The most common process in the United States for removing lignin is kraft pulping, which involves cooking wood chips with a mixture of sodium hydroxide and sodium sulfide to partially depolymerize and solubilize the lignin. The resulting kraft pulp has a value six to eight times that of the original wood raw material. Most of the byproduct kraft lignin is burned to provide fuel for pulp and paper mills and to recover and regenerate the inorganic pulping reagents. The most modern mills actually produce more energy than they consume, reducing the need for fossil fuels. About 0.1 percent of kraft lignin produced in the United States (45,000 tons) is diverted for use as a chemical resource, and more could be diverted if its value in the production of chemicals, building materials, and other products exceeded its value as a fuel.

The other major chemical pulping process, used more in Europe and elsewhere outside the United States, involves cooking wood with sulfite salts to convert the lignin to water-soluble ligninsulfonate. Although some such mills still operate in the United States, mills must absorb high costs to recover waste materials that were previously disposed of in streams and waterways. Unlike the kraft process that degrades carbohydrates that are inadvertently solubilized during pulping, sulfite pulping produces spent liquor that contains nondegraded fermentable monomeric and oligomeric sugars. Some mills in the United States and Europe ferment sugars in the spent liquor to produce ethanol.

Acid hydrolysis of wood is an old technology developed extensively during and following World War II. It involves using dilute or concentrated sulfuric acid to effect hydrolysis at temperatures between 120° and 170°C. The solid residue resulting from hydrolysis contains some of the cellulose and most of the initial lignin. Notable processes in the published literature that have not, however, been adopted in wide-scale commercial practice include the "Hokkaido Process" (Bungay, 1981), the "Berguis Rheinau Process" (Bungay, 1981), and continuous acid hydrolysis. Hydrochloric acid has also been used as a lignocellulose pretreatment, generally in concentrations of 35 to 50 percent and at temperatures between 100°C and 170°C. O'Neil (1978) has reviewed variations in the hydrochloric acid hydrolysis process.

Potential

Lignin is an important coproduct with sugars resulting from the hydrolysis of lignocellulosics. The overall economics of biorefineries that use lignocellulosics as raw materials would be improved by finding uses for lignin and avoiding lignin disposal costs. At a minimum, lignin can be burned to produce electricity and steam for supplying the biorefinery, with excess electricity exported to the grid. Higher-value uses may be possible, however.

Kraft lignin and ligninsulfonate have a wide variety of relatively low-value applications as polymers. Attempts by pulp and paper research laboratories to develop additional markets for byproduct lignins have failed. Similarly, although low-molecular-weight compounds, particularly phenols, can readily be produced from kraft lignin, they have not been produced in a commercially competitive manner. However, a recent study by the U.S. Department of Energy (DOE) concluded that phenolics and anthraquinones could be produced competitively from lignin by pyrolysis of lignocellulosics (Bozell and Landucci, 1993). The study further concluded that production of benzene, toluene, and xylenes from lignin by pyrolysis also looked promising.

Although carbohydrates comprise about 70 percent by dry weight of lignocellulosic plant material, most are in a form that cannot be readily fermented. Without pretreatment—chemical, mechanical, biological, or combined—carbohydrate yield following hydrolysis will be low, making processes based on fermenting these sugars less economical. Purdue University researchers undertook a major effort in the late 1970s using various solvents to dissolve and swell lignocellulose (e.g., Ladisch et al., 1978). The key to their approach was using a solvent that could permeate the lignocellulose and disorganize its crystalline structure. This change increases lignocellulose reactivity and makes it more amenable to subsequent hydrolysis. One method causes the dissolution and swelling of lignocellulose using ethylenediamine (25 to 30 percent) in water plus cadmium oxide (4.5 to 5.2 percent) at room temperature. The treatment makes the lignocellulose highly susceptible to enzymatic hydrolysis and enables nearly 100 percent of the theoretical glucose yields. This type of process is in experimental stages and has not yet been adopted by industry.

Many solvents have been shown to extract the lignin effectively from lignocellulose, including organic solvents such as butanol and inorganic extractants such as ammonia. The use of solvents to delignify wood and make the lignocellulosic biomass more susceptible to biological processing is often called the "organosolv" process (Holtzapple and Humphrey, 1984). Much work has been done on this process, and a demonstration pulp mill based on the process (Repap) currently operates in Canada. Avgerinos and Wang (1983) also have reported the use of selective solvents for delignification of lignocellulose.

A novel pretreatment for lignocellulosic biomass, known as the ammonia fiber explosion (AFEX) process, has been developed for various feedstocks, including alfalfa hay, bagasse, wheat straw, switchgrass, and coastal bermudagrass (Holtzapple et al., 1991; de la Rosa et al., 1994). This technology uses liquid ammonia at elevated pressures and temperatures (between 50°C and 90°C) to permeate the lignocellulosic biomass (Dale and Moreira, 1982). The treated material is then decompressed to achieve an "explosion" of the biomass, yielding material that is much more susceptible to microbiological and enzymatic degradation. The AFEX process followed by treatment with cellulases and hemicellulases gives high sugar yields (approximately 90 percent of theoretical estimates) from various raw materials using low enzyme levels (about 5 IU per gram of dry biomass; Holtzapple et al., 1991; de la Rosa et al., 1994). More than 99 percent of the ammonia in the AFEX process can be recovered and reused. In comparison to the many pretreatment technologies studied over the past 80 years, the AFEX process appears to be attractive from technical, environmental, and economic perspectives, but it is too early in the development stage to make accurate judgments on its potential commercialization.

Another relatively new and promising lignocellulose pretreatment involves using liquid water under pressure at temperatures over 200°C (Van Walsum et al., 1996). While data on this approach are still incomplete, the possibility of using liquid water as a pretreatment chemical has obvious advantages. Pressurization of wood with steam to 45 to 52 atmospheres with subsequent decompression shatters the structure of lignocellulose (Bungay, 1981). This method has been shown to be effective with various biomass materials and forms the basis of several processes to disintegrate the lignocellulosic biomass (Carrasco et al., 1994). The technology has undergone pilot-scale development but is not yet used in industrial processes.

Conversion of lignocellulosic waste materials is an active area of research. The conventional technology for converting cellulose from waste paper to glucose is a stirred-tank batch reactor with cellulase enzymes; however, the enzyme hydrolysis takes over 40 hours to complete and the enzyme costs are prohibitive, at least at the high enzyme levels typically used (>20 IU per gram of dry biomass; Scott et al., 1994). An advanced alternative reactor technology couples two types of continuous bioreactors in a stirred-tank attrition bioreactor containing cellulase enzymes, followed by a fixed-bed reactor with immobilized cellobiase (cellobiose hydrolase) entrapped in gel beads. Under experimental conditions, this approach achieved essentially complete hydrolysis of waste paper in 25 hours (Scott et al., 1994). Since waste paper is already substantially pretreated, this process may not prove as effective for untreated lignocellulosic substrates such as aspen or poplar. A roadblock to lignocellulose conversion is the high cost of cellulase enzymes and lack of proven alternatives (Rooney, 1998).

Thermal, Chemical, and Mechanical Processes

Certain thermal, chemical, and mechanical processes for the conversion of renewable resources into chemicals and fuels have been practiced for many years (for further details see Bungay, 1981, and Elliott et al., 1991). Others have been experimentally or theoretically demonstrated but are not yet in commercial use. In general, the role of mechanical processes is to prepare various feedstocks for subsequent conversion by other types of processes. This section describes proven and potential thermal, chemical, and mechanical processes. It does not cover pulping technologies because pulp cost is generally too high for most uses in biobased products.

Proven

Direct combustion of wood for cooking and heating has been practiced for centuries. Industrial-scale generation of steam that, in turn, can be used to produce electricity frequently relies on combustion of wood biomass and municipal solid wastes. The biomass power industry in the United States is expected to grow from 31 billion kilowatt hours in 1990 to 59 billion kilowatt hours in 2010. This is approximately 1 percent of the total generation of electricity in the United States (EIA, 1997), including conventional (e.g., hydroelectric) and renewable resources.

Biomass from renewable resources often must be mechanically densified before it provides an effective combustion fuel. Densification of woody biomass, for example, generally involves pelting, cubing, briquetting, extrusion, and roller compression. A number of densifiers are commercially available for industrial operations (Bungay, 1981).

Mechanical and thermal processes, such as distillation, are involved in the refining of pine tree exudates into industrial chemicals (e.g., turpentine, oleoresins, gum spirits, pinenes, camphene, abietic acid). The exudates, or "naval gum stores," are byproducts of the paper pulp and lumber industries, and most naval gum currently comes from processing of kraft pulp. The naval gum industry, although not large, is economically viable in combination with the lumber and paper industries.

Pyrolysis is a well-established thermochemical technology using "destructive distillation" to convert biomass into useful chemicals and fuels (see Elliott et al., 1991, for review). High temperature and limited air—sometimes in the presence of a catalyst—yield primarily hydrocarbon liquids (equaling 50 to 70 percent of inlet raw material), gases, and char. Major product yields following liquid pyrolysis can vary from 25 to 70 percent of inlet raw materials.

Commercial production of methanol has been accomplished by the destructive distillation of wood. Despite earlier optimistic projections, today no large-scale methanol manufacturing plant relies on destructive distillation (Hokanson and Katzen, 1978; Stinson, 1979). Most methanol is produced by the lower-cost method of chemically oxidizing natural gas (methane).

The production of furfural (the chemical furaldehyde) and its derivatives is another well-established technology (McKillip and Sherman, 1979). The major raw materials are hemicelluloses from annual crops. Although any agricultural product containing high-pentose-content hemicelluloses could be used in theory, only a few (corn cobs, oat hulls, rice hulls, and bagasse) are available in tonnage quantities within economic hauling distance of furfural production plants. Furfural production involves loading raw materials into a digester containing a strong inorganic acid. The hemicelluloses are hydrolyzed to pentoses, and these pentoses are then cyclodehydrated to furfural. High-pressure steam provides the necessary temperature, and the furfural is steam distilled. Various subsequent chemical modifications can produce furan derivatives. The current market for furfural is small; however, furfural was the sole source of adipic acid and hexamethylenediamine in the early days of nylon 66 and may have renewed potential in the future.

Well-developed chemical processes exist for deriving cellulose esters from natural cellulose. Cellulose acetate is the major cellulose ester currently in commercial production. The production process uses sulfuric acid as a catalyst with acetic anhydride in an acetic acid solvent. Cellulose acetate production declined from 392,000 tons in 1979 to 321,000 tons in 1993, mostly because other organic polymers have partially replaced cellulose acetate.

Cellulose films can be made by dissolution of very clean cellulose preparations in solvents. This process has been used for several decades to produce cellophane, the basis for camera films and hot dog casings. However, cellophane at present is not competitively priced for many applications, largely because the dissolution process for cellulose xanthate requires carbon disulfide, an expensive solvent that poses environmental hazards. New solvent-based processes could increase the competitiveness of cellulose films and enable replacement of petrochemical-derived plastics in applications such as packaging. Less expensive dissolution processes also could make rayon and other cellulose polymers more competitively priced. Two companies from the United Kingdom and the Netherlands recently agreed to take this approach and formed a joint venture to develop and produce a solvent-spun cellulose textile yarn.

Gasification is an established technology involving reaction of carbonaceous materials with steam and oxygen to produce a mixture of carbon monoxide, hydrogen, carbon dioxide, methane, tar, and char. Numerous different reactors have been used for gasification, and coal is usually the raw material (Bliss and Blake, 1977). Only limited commercial development of gasification using renewable raw materials, such as wood, to produce useful products has occurred, mainly because of the poor economics.

Fractionation of protein-rich renewable resources (i.e., forages and grasses) is a thermal/mechanical process that has been proposed for producing animal feeds. Pirie and co-workers reported the first experiments producing "leaf protein" feeds in the 1940s (Pirie, 1971). Despite much subsequent research, a viable industry has not yet emerged. This failure is due at least in part to economic considerations. The existing leaf protein processes do not increase the value of the fibrous portion of the forage or grass (i.e., the cellulose, hemicellulose, and lignin) that constitutes over 80 percent of the plant material's dry weight. The economics of protein extraction might be more favorable if the fibrous portion's value could be enhanced—for example, by fermenting the carbohydrates to fuels and chemicals.

Potential

Liquefaction of wood is a theoretically possible technology that produces synthetic fuels having properties similar to petroleum and natural gas. However, past work on synthetic fuel technologies has focused primarily on coal as the raw material, and wood liquefaction differs from that of coal. The wood's lignocellulosic biomass cannot be catalytically hydrogenated to produce liquid fuels because liquefaction of cellulose requires removal of oxygen as well as addition of hydrogen. In several feasibility and pilot studies, researchers have catalytically removed oxygen from lignocellulosic biomass and then used liquefaction and catalytic hydrogenation to produce liquid fuels (Bliss and Blake, 1977; Elliot and Walkup, 1977). More recent studies using extrusion-fed reactors have obtained 80 to 100 percent of the theoretical oil yield from wood flour (White and Wolf, 1988). Other liquefaction approaches (by solvolysis, catalysis, steam/water, or high pressure) have generally given lower oil yields, typically 40 to 60 percent (Bain, 1993).

Another possible approach builds on earlier protein fractionation work to coproduce protein and ethanol (Dale, 1983; de la Rosa et al., 1994). The process combines protein extraction from dry forages with hydrolysis of the extracted fiber to fermentable sugars. Researchers pretreated coastal bermuda grass with the ammonia fiber explosion process (AFEX) and then extracted more than 80 percent of the protein with dilute calcium hydroxide. Hydrolysis of the resulting extracted fiber yielded more than 90 percent of the theoretical sugar yield, which was subsequently fermented to ethanol (de la Rosa et al., 1994). Coproducing protein and fuels could help reduce the perceived or real conflicts inherent in allocating renewable resources to the two important human needs: food and fuel.

Biological Processes

Biological processes often play a role in the pretreatment and conversion of renewable resources into industrial products. For example, starch resulting from corn wet milling is enzymatically hydrolyzed to glucose syrup that is then fermented to ethanol and other products. In bioprocesses, fermentation and enzymatic reactions carry out most classes of reactions used in the chemical industry, including oxidations, reductions, hydrolysis, esterifications, and halogenations. Unlike thermal and chemical processes, bioprocesses occur under mild reaction conditions, usually result in stereospecific conversions, and produce a smaller number of generally nontoxic byproducts (see Box 4-1). One drawback is that bioprocesses typically yield dilute aqueous product streams, requiring subsequent thermal, chemical, or mechanical processing for separation and purification. Other barriers to economic competitiveness include feedstock cost, product yield, and high capital costs.

Proven

The fermentation industry produces alcohol, organic acids, polymers, antibiotics, amino acids, enzymes, and other products. The most common production technology is the conventional stirred-tank batch fermentor. However, continuous stirred-tank reactors with process control can enhance the economics of fermentation by allowing higher volumetric productivity. Immobilized cell reactors can provide a higher productivity in the fermentation broth than do stirred-tank reactors, but difficulties can arise in ensuring that adequate carbohydrate is transferred to the fermenting cells ("mass transfer") or in removing heat produced by the biological reactions.

> **BOX 4-1**
> **Softening Wood the Natural Way**
>
> A new experimental process called "biopulping" uses natural lignin-degrading fungi to "soften" wood prior to pulping. Wood usually is pulped either chemically, by modifying and dissolving the lignin, or mechanically, by grinding to separate the fibers. Studies of biopulping have generally examined it as an "add on" to mechanical processes, although it shows promise in combination with chemical processes as well. Biopulping decreases the energy required for mechanical pulping by 30 to 50 percent, results in stronger paper, and is environmentally sound. Electrical energy is the major cost associated with mechanical pulping, and an energy savings of even 25 percent by addition of the fungal process would justify substantial investment in the latter method.
>
> Research on biopulping has resulted from an eight-year collaboration among the U.S. Department of Agriculture (USDA) Forest Service's Forest Products Laboratory, the University of Wisconsin's Biotechnology Center, the University of Minnesota, and industry. Only a few of the several hundred fungal species and strains examined thus far show promise; the relatively uncommon *Ceriporiopsis subvermispora* appears to be superior. This fungus species is effective with aspen and loblolly pine, two woods used extensively in mechanical pulp production. Biopulping requires about 2 weeks and causes little damage to cellulose. Brief steaming of the wood chips enhances the process because it gives the fungus a competitive advantage over indigenous microorganisms, and supplementing the fungal mycelium with corn steep liquor reduces the quantity (cost) of the initial fungal inoculum. Modest rates of aeration can control the self-heating associated with biopulping.
>
> As presently envisioned, commercial biopulping would involve steaming, cooling, and inoculating the wood chips and then allowing incubation with aeration for about two weeks prior to pulping. A disadvantage of biopulping is that it darkens the chips, requiring increased chemical use during pulp bleaching. An advantage is that biopulping largely destroys pitch components (lipophelic constituents of wood, e.g., oils, resins, fatty acids) that cause problems during the pulping process.
>
> A pitch reduction process similar to biopulping has already been commercialized. The fungus in this case, *Ophiostoma pilferum*, does not degrade lignin. When inoculated into wood chips being piled for storage, the fungus selectively degrades the pitch. A white strain of *O. pilferum* used in a product from the Sandoz Chemicals Corporation also acts as a biocontrol agent, keeping out dark-colored fungi that normally discolor wood chips during storage.
>
> SOURCE: Kirk et al. (1993).

Fuel ethanol production involves fermentation of glucose syrup by the yeast *Saccharomyces cerevisiae* in large-scale, high-capacity fermentors. The technology has evolved from a batch-scale system to the cascade system now found in almost all large U.S. plants (Hacking, 1986). The cascade system approximates a continuous operation by continuous flow through a series of fermentors.

The major organic acids produced by fermentation are citric and acetic acid for food industry uses. Other organic acids, primarily gluconic and lactic acids, are produced in smaller quantities. Production of citric acid worldwide relies on either the yeast *Candida lipolytica* or the fungus *Aspergillus niger*. Today, citric acid production usually involves submerged aerobic fermentations using glucose or molasses as the carbon source (USDA, 1994). Over the past 30 years, mutation and selection programs have yielded superior production strains of the yeast and fungus that can produce high concentrations of citric acid with few or no undesirable byproducts, such as isocitric acid.

Research on acetic acid production has focused for more than 50 years on homoacetogenic bacterium, such as *Clostridium thermoaceticum*. ("Homoacetogenic" refers to organisms that ferment a variety of substrates to primarily acetic acid.) The bacteria can convert glucose, xylose, and some other pentoses almost completely to acetate, which is formed by neutralization of acetic acid with dolomite. The energy crisis of 1972 stimulated general interest in the complex metabolism of homoacetogenic bacteria, and the U.S. Department of Transportation's Federal Highway Administration has supported recent related research. Calcium magnesium acetate is important as an organic, biodegradable, noncorrosive deicing salt and as an additive to coal-fired combustion units for control of sulfur emissions.

The biopolymer xanthan gum is produced by large-volume fermentation of molasses or corn syrup using the aerobic bacterium *Xanthamonas campestris*. This polysaccharide has uses in the food, paper, and oil field industries because of its high viscosity at low concentrations. The high viscosity also, however, makes heat and mass transfer problematic, especially in large-scale fermentors.

Large-capacity fermentors, typically stirred tanks, are used in the production of amino acids for food enhancers and animal feed supplements. The Japanese fermentation industry has dominated worldwide production of amino acids for the past 20 years and during this period has used random mutation and selection programs to develop high-producing microbe strains for many amino acids (e.g., glutamic acid, phenylalanine and lysine). ADM has recently begun to challenge dominance of the amino acid industry by Japanese companies (see Box 4-2). This example illustrates how the combination of low-cost raw materials and advanced technologies potentially makes U.S. companies competitive, even in industries historically dominated by foreign companies.

Microbial fermentation can produce enzymes for a variety of commercial processes and products, including production of corn sweeteners, laundry detergents and digestive aids, cheese and beverages, leather and textiles, and medicine and clinical diagnostics. Major industrial uses of

> **BOX 4-2**
> **The Changing U.S. Role in Worldwide Amino Acid Production**
>
> Industrial production of amino acids dates back to the 1950s, when the major amino acid product was sodium glutamate—a food taste enhancer. The United States and several Asian countries were the major manufacturers, producing glutamic acid by hydrolysis of wheat gluten.
>
> In the late 1950s, Japanese scientists invented new methods to produce glutamic acid based on direct fermentation by selected microorganisms. The development of random mutagenesis and auxotrophic mutants made it possible to isolate production microbe strains that yielded particularly high levels of glutamic acid. Further application of these technologies, along with the development of suitable microorganisms, allowed Japanese companies to produce numerous amino acids by fermentation processes. Their eventual dominance of the amino acid industry included numerous amino acids other than glutamic acid, such as lysine, tryptophan, phenylalanine, threonine, and aspartic acid for human nutrition and animal feeds. The last of the early U.S. producers of amino acids, Stauffer Chemical Company, halted production in the 1970s.
>
> Archer Daniels Midland Corporation (ADM), a U.S. processor of agricultural commodities, challenged the dominance of the amino acid industry by Japanese companies in the 1990s. Amino acid products are large volume and low value compared to other fermentation products. ADM's strategy recognized that raw material and energy costs as well as efficient waste disposal systems are the keys to economic success. ADM built a 60,000-ton-per-year lysine plant in Decatur, Illinois, that incorporated sound engineering and the company's previous experience producing commodity products (starch, syrups and ethanol). Today, ADM is a major player in the worldwide amino acid market and has successfully relaunched a U.S. biobased industry by capitalizing on the nation's rich agricultural base.
>
> Source: Prepared by Dr. Greg Zeikus.

enzymes occur in the liquefaction of starch to produce maltodextrins, corn syrup, high-fructose corn syrup, and laundry detergents, and as catalysts in the manufacture of other commercial products. Enzymes are used to catalyze reactions of commercial interest where chemical catalysts fail or perform poorly. Enzymes are widely used in the food, chemical, and pharmaceutical industries because of their unusually high catalytic activity, selective action, ability to function under mild conditions, frequent low cost, biodegradability, and high product yields.

Enzyme-catalyzed processes can be accomplished in both stirred-tank reactors and immobilized enzyme reactors. The corn-ethanol industry relies on enzymatic digestion of starch in stirred-tank reactors to produce glucose. Two processes are involved: digestion by amylases in a liquefaction process and digestion by amyloglucosidases in a saccharification process. The amylases now used for liquefaction are thermostable; some come from genetically engineered microorganisms. Ongoing research is examining ways to improve the liquefaction and saccharification steps by using immobilized enzyme reactors (e.g., Zanin et al., 1994; Gerhartz, 1990). Immobilized-enzyme reactor technology contributed to the greatest success in enzymatic processing to date—the commercial production of high-fructose corn syrup from glucose using the enzyme glucose isomerase (Klibanov, 1983). The example of high-fructose corn syrup demonstrates how advanced bioreactor concepts derived from research and development can be commercialized (see Box 4-3).

Several other significant examples exist of commercial enzymatic processing (Klibanov, 1983). Penicillin G and V catalyzed by penicillin acylase produces the 6-aminopenicillanic acid used in the synthesis of semisynthetic penicillin. The production of L-malic acid, an acidulant used in beverages, involves conversion of fumaric acid by fumarase. A number of different L-amino acids used in food and feed supplements are produced enzymatically. Finally, acrylonitrile catalyzed by nitrile hydratase yields acrylamide, a chemical used to make polymers.

In recent years new opportunities for enzymatic processing have arisen from changing market conditions and growing environmental concerns. For example, in the pulp and paper industry, enzymes can help reduce chlorine use in the bleaching process. The kraft process is the major pulping process in the United States. It produces cellulose pulp that is brown because it contains residual modified lignin. The lignin is whitened by oxidative degradation and washed using chlorine and chlorine dioxide. However, environmental concerns have begun to reduce chlorine use. Work in the late 1980s showed that hemicellulases, primarily xylanases, could facilitate chlorine bleaching and thereby reduce bleaching chemical requirements and the production of chlorinated waste products. The hemicellulases are thought to "uncover" lignin shielded by residual hemicelluloses. Scandinavian and Canadian mills now use xylanases at a commercial scale, and other mills are examining this approach (Daneault et al., 1994), along with other technologies. Oxidoreductase enzymes (e.g., laccase) also have potential value in these applications. Nonchlorine bleaching agents, such as ozone, oxygen/alkali, and peroxide, are gradually replacing chlorine, and xylanases will probably prove useful in combination with these chlorine substitutes (Allison and Clark, 1994).

The pulp and paper industry has begun to adopt several other new enzyme applications. Pitch reduction by mechanical pulping with lipases is now used commercially in Japan (Fujita et al., 1992). In France, cellulase/hemicellulase mixtures are used to improve drainage of water from pulp in paper making. Cellulase/hemicellulase mixtures are also effec-

> **BOX 4-3**
> **Making Alternative Sweeteners from Corn**
>
> Sucrose, the constituent of table sugar, has always been part of the human diet. Beets and sugar cane are the major sources of sucrose. U.S. production of these crops is limited, however, because appropriate climatic conditions occur only in certain regions of the country. In contrast, the United States is a major producer of glucose from starch. Glucose and fructose are the two monosaccharides that make up sucrose. Unfortunately, it is the fructose and not the glucose that makes sucrose sweet.
>
> Volatility of the world sugar market has created an incentive to seek domestic replacements for this commodity food ingredient. In the mid-1950s, U.S. scientists discovered an enzyme that could convert glucose into fructose (fructose is an isomer of glucose). The enzyme was developed in other countries and then later returned to the United States under a licensing agreement. Its discovery eventually led to growth of the high-fructose corn syrup industry, which uses abundant corn starch from the U.S. agricultural sector as its starting raw material.
>
> A number of U.S. companies now produce high-fructose corn syrup, including A.E. Staley, Roguette, ADM, CPC International, and Cargill. It is truly a biobased industry: the raw material is corn starch, and the conversion processes are entirely enzymatic (involving amylase, amyloglucosidase, and glucose isomerase). The high-fructose corn syrup industry has demonstrated that immobilized enzyme processing and chromatographic separation of biomass can be efficient and economical on a large scale. Since the start of the industry in 1970, total annual domestic production of high-fructose corn syrup has increased to more than 15 billion pounds. This development has added value to a commodity agricultural product, lessened U.S. sugar imports, and aided the growth of the grain ethanol fuel industry. It should be noted that the development of high-fructose corn syrup has been aided by the high cost of sugar arising from the sugar import quota.
>
> ---
>
> Source: Prepared by Dr. Donald Johnson.

tive in deinking office waste paper (Jeffries et al., 1994), and this method is also likely to come into commercial use.

Potential

Funding by government agencies (DOE, USDA, National Science Foundation) from 1975 through 1985 supported fundamental research in biological technologies for ethanol production (NRC, 1992). A significant focus of this research was improved cell suspension and immobilized-cell reactor designs (Maiorella et al., 1981 and 1984; Bailey and Ollis, 1986). One promising approach resulting from this work provides an alternative to cell suspension bioreactors—a fluidized bed using microorganisms at high concentrations immobilized in either cross-linked industrial grade carrageenan, cross-linked modified bone gel, or other porous materials (Scott et al., 1994; Davison and Scott, 1988). A fluidized-bed approach using the bacterium *Zymomonas mobilis* with industrial feed materials has a volumetric productivity at least 10 times greater than conventional (batch reactor) technologies (Webb et al., 1995). Ethanol product concentrations of over 7.5 percent (w/v—weight of ethanol per volume of solution) can be maintained at an overall volumetric productivity of more than 60 grams ethanol per liter per hour.

Another advanced bioreactor concept, simultaneous saccharification and fermentation (SSF) reactors, can produce ethanol from mixed waste papers, agricultural wastes, and pulp and paper mill wastes (Emert and Katzen, 1980, Katzen and Fowler, 1994). The SSF reactor combines enzymatic and fermentation steps in one process unit using novel recombinant strains of bacteria. In-house production of cellulase enzymes (to further reduce costs) will be used in a planned application of this technology to convert waste paper in Florida (Katzen and Fowler, 1994).

A great deal of interest centers on use of the bacterium *Clostridium acetobutylicum* to produce acetone and butanol from starch, especially if the fermentation can be performed on a continuous scale. This fermentation has been practiced for more than a century at a batch scale, but the low-cost availability of petrochemicals has limited commercial production of bulk quantities by fermentation. As a group, these bacteria can grow on a wide variety of compounds, including waste streams that otherwise present disposal problems, such as cheese whey, fruit pulp, and partially degraded hemicelluloses. The fermentation produces large volumes of carbon dioxide and hydrogen as byproducts, and efficient recovery of these gases increases the process's value. Another "byproduct" is the bacterial biomass, a potential source of animal feed.

Genetic engineering and improved continuous production technology potentially could make fermentation by clostridia economically competitive with petrochemical sources. However, these advances pose significant challenges. Engineering the bacteria to produce a single end product from fermentation of carbon-rich compounds, for example, is made difficult by the inherent complexity of carbon flow in the clostridia. Carbon enters these bacterial cells in many forms. Simple and complex carbohydrates are readily used, and carbon is channeled by intermediary metabolism from pyruvate through acetyl-CoA to ethanol, acetate, acetone, butyrate, and butanol. All of these end products are produced to some extent, but some species may make predominantly one or more of them. The relative production of certain products depends on growing conditions such as carbon source, pH, and temperature. For instance, high glucose concentrations cause high solvent production and low acid production. Conversely, cells growing under conditions of limited carbon produce only acids because they do not enter the solvent production stage.

Additional problems arise because the clostridia and other microorganisms that make desirable products come from extreme environments. These "extremophiles" not only survive but actually thrive under conditions such as high temperatures (>100°C), high salinity, acidic or alkaline solutions, or where the atmosphere lacks oxygen (Gilmour, 1990). The bacteria have evolved special enzymes and metabolic strategies that allow growth in hostile conditions. They may survive low oxygen levels, for example, by channeling or adjusting the way they convert various intracellular metabolites. Consequently, attempting to direct carbon flow to desired products often fails to achieve the intended result.

Experimental attempts to drive carbon into one biosynthetic pathway over another by inducing overexpression of an enzyme may not succeed because the availability of cofactors and substrates becomes limiting. In theory, genetic engineering approaches could overcome this problem by directly modifying amounts of cofactors and substrates. In addition, requirements for growing clostridia under extreme environmental conditions and dealing with production of flammable hydrogen gas as a byproduct present technical challenges in process design and operation. Another approach has been to clone genes from clostridia for solvent production into other bacteria such as *Escherichia coli* or *Bacillus*. This strategy has had only limited success so far because these organisms, unlike clostridia, have not evolved to tolerate high levels of solvent during growth.

NEEDED DEVELOPMENTS IN PROCESSING TECHNOLOGY

The establishment of large new biorefineries will require feedstocks of satisfactory price, quality, and quantity that are consistently and reliably available for processing. Sufficient numbers of trained individuals must be available to design, build, and operate these refineries, and economical and effective processes for converting raw materials into value-added products must be known or reasonably well developed. This section identifies key areas of processing technology where focused research and development efforts would have the greatest impact.

Upstream Processes

Renewable natural resources, including lignocellulosic biomass, will provide the raw materials for future biobased industries. Developing economical processes for pretreating this biomass will be essential to make sufficient carbohydrate polymers available for subsequent biological conversions. An efficient approach to this task would be to test, at a demonstration scale, the most promising pretreatment technologies developed over the past 10 to 15 years in order to assess their large-scale technical and economic feasibility.

Extensive research has examined enzymatic degradation of cellulose and hemicellulose into fermentable sugars (NRC, 1992). Improvements in pretreatment technology (to enhance enzyme accessibility) and enzyme hydrolysis would lower the economic barriers to commercial lignocellulose conversion. Developments that reduce enzyme costs and increase the efficiency of xylose fermentation would also improve the overall economics. A particularly attractive possibility is to combine, in a single microorganism, the ability to produce extracellular hydrolytic enzymes that degrade cellulose and hemicelluloses with fermentation machinery and thereby directly convert biomass to useful products.

Bioprocesses

Advances in bioprocess engineering will play a key role in enabling the development of biorefineries. Advanced bioreactor concepts, especially for high-volume products, might allow significant increases in productivity and thus reduce capital and operating costs. The goals for future research on fermentation operations should include:

- combining the biological and physical operations of sugar production, fermentation, and product recovery in fewer vessels and fewer microorganisms to reduce capital costs and inhibition by microbial products, thereby increasing rates, yields, and selectivities.
- improving bioreactors for heat, momentum, and mass transfer for viscous, non-Newtonian fermentation broths and solid-liquid broths;
- developing new methods for monitoring biological processes, such as discrete sensors using microfabrication, real-time monitoring, and digital imaging of bioreactors; and
- developing new concepts in process control, such as the application of expert systems, artificial intelligence, neural networks, and principal component analysis.

The development and improvement of biological conversion processes will depend on continuing research to elucidate fundamental biological principles, combined with engineering analyses to improve biobased product yield (raw material conversion), selectivity (reduction of byproducts), and productivity (enhanced rate of production). Future work on molecular biology, genetics, and microbial physiology should include:

- identification of genetic changes that enable use of multiple substrates;
- analysis of biochemical pathways combined with directed mutagenesis (deletion and addition of enzymes) to reduce production of undesirable byproducts; and
- analysis of processes or steps in biochemical pathways that limit formation of desired products and identification of genetic changes to alleviate these limitations.

Microbiological Systems

Microbiological systems, particularly those involving anaerobic bacteria, are essentially multicatalytic reactors that direct the microbial metabolism toward the production of useful biobased products. Effective control of the extracellular environment by engineers would allow optimum product synthesis. Such control, however, will require a good understanding of the complex intracellular biochemical reactions involved in cellular metabolism and physiology. Several groups of bacteria in particular—the anaerobic bacteria, the Archei bacteria, and the extremophiles—provide a vast potential of host organisms for the production of chemicals and fuels from renewable resources, yet they have received relatively little research investment in comparison to aerobic microorganisms. Capturing their potential will require both research on fundamental aspects of microbial metabolism and physiology as well as tools for assessing and manipulating such processes. For example, one important goal is to identify strategies for either improving biosynthesis of products in bacteria that normally produce them or moving the relevant genes from these bacteria into new hosts. In either case, a greater understanding of the basic biology of biosynthesis and regulation of these compounds will be necessary. Future research on microbial metabolism, nutrition, and physiology should include:

- Development of basic tools to assess metabolic pathways for microorganisms such as intracellular measurements of metabolic intermediates.
- Analyses of biochemical pathways that integrate these basic intracellular measurements. Such analyses will provide fundamental understanding of the microbial metabolism and physiology necessary to focus metabolic engineering manipulations on enhancing organisms' overall productivity.

- Identification of microorganism strains that can produce products for important biobased markets (see also Chapter 3), including organisms that can ferment 5-carbon and 6-carbon sugars to ethanol, organisms that can efficiently use sugar syrups from lignocellulose to produce other target chemicals (e.g., butanol and acetone), and organisms that can use ethanol and other oxygenated intermediates as a substrate to produce target chemicals.
- Understanding the interactions between the environment and the metabolism and physiology of microorganisms, particularly unusual microorganisms such as Archei bacteria and extreme thermophiles. Such understanding will enable manipulation of bioreactor environments to mimic the natural environments in which these organisms have evolved.
- Studies of the fundamental principles of microbial physiology that affect regulation of metabolic pathways. Such information may make it possible to enhance product synthesis and secretion by overcoming biochemical regulatory constraints and metabolic flow bottlenecks.
- Basic research on principles of intermediate microbial metabolism to gain a better understanding of how concentrations of substrate or product can inhibit rates of product formation. Such understanding will aid in engineering bioreactor control to enhance the rate and conversion of raw materials into useful products.
- Basic research on cultures containing multiple microorganisms, particularly those in which the microorganisms have symbiotic relationships that increase growth and product formation. Advanced analytical tools are needed to assess the metabolism and physiology of these mixed-culture systems and the intricate relationships among mixed-culture populations.

Enzymes

The scope of current industrial applications of enzymes is limited by certain drawbacks of enzymes as practical catalysts. Such drawbacks include the insufficient operational stability of many enzymes, their real or perceived requirement for an aqueous reaction medium, and their often very narrow substrate specificity (i.e., enzymes frequently catalyze a particular reaction with a given substance but not an analog that is of commercial interest). Recent scientific advances in enzymology and molecular biology promise to eliminate, or at least alleviate, these limitations and significantly expand the opportunities for enzymatic processing.

For example, genetic engineering now makes it possible to replace any amino acid residue in an enzyme with any other amino acid virtually at will. Since most properties of enzymes are determined by their amino acid sequence, such replacement should in principle bring about a desired change in enzyme performance, provided that the structure-function relationship is understood. This type of "protein engineering" has successfully enhanced the stability of enzymes against heat and oxidizing agents as well as broadened their substrate specificity (Knowles, 1987). However, the greatest impacts of protein engineering may well lie ahead.

Other major developments have made it clear that many enzymes can work as catalysts not only in aqueous reaction media but also in organic solvents to efficiently convert water-insoluble substances and catalyze new reactions (Klibanov, 1990). Also, the solvent can control enzyme properties, such as selectivity of action (Wescott and Klibanov, 1994). An industrial example of this nonaqueous bioprocessing is lipase-catalyzed conversion of palm oil to cocoa butter substitutes developed by the Unilever Company in Europe (Macray, 1985). It may even be possible to create artificial enzymes with tailor-made properties—for example, by generating "catalytic antibodies" (Schultz et al., 1990). Such advances should further increase the versatility of enzymes and the applicability of enzymatic catalysis to a broader range of industrial processes.

Professor Stephen Withers of the University of British Columbia has altered, by genetic engineering, the active site of a xylosidase so that it can no longer catalyze the hydrolysis of xylosidic bonds, its normal reaction, but only their synthesis (Withers et al., 1996). In the new enzyme the normal glycosyl-enzyme intermediate is intercepted with a sugar acceptor, rather than with water, which gives hydrolysis, forming a new glycosidic bond. This initial step in reversing enzyme activity gives hope that it can be done with other enzymes, but extensive fundamental research to elucidate the precise structure and mechanism of the catalytic sites of the enzymes in question must precede such work.

To capitalize on and expand recent developments in enzyme processing, future research in this area should:

- further explore catalysis by enzymes from thermophiles and other extremophiles;
- elucidate the mechanisms that allow enzymes isolated from extremophiles to remain stable in hostile environments (such mechanisms might eventually be incorporated into stabilization strategies for industrial enzymes from nonextremophiles);
- broaden the substrate specificity of enzymes by means of site-directed mutagenesis;
- identify enzymes that can produce products for important bio-based markets (see also Chapter 3), such as enzymes that can use

ethanol and other oxygenated intermediates as a substrate to produce target chemicals;
- enhance and develop rational control of enzyme performance in organic solvents;
- investigate catalytic antibodies and nonprotein enzyme models to carry out industrially relevant conversions of renewable feedstocks; and
- improve inexpensive large-scale production of industrially useful enzymes.

Downstream Processes

A major weakness of fermentation processes in the production of large-volume, low-value chemicals is that the products are in dilute levels in aqueous effluent streams. In the production of organic solvents, this low product concentration reflects the microbe's low tolerance for these solvents. For organic acids the protonated species of the acids adversely affect the biosynthetic enzymes; neutralization of the acid is often required during fermentation.

Conventional unit operations—such as distillation, adsorption, liquid-liquid extraction, pervaporation, ion exchange, crystallization, and membrane separation—have been used for downstream purification and separation of fermentation-derived products. However, these separation technologies were developed for other systems, usually petroleum refining, and applying them to fermentation systems in a cost-effective manner will require much research. New concepts and methods must be developed to address the unique problems of fermentation products in aqueous solution, as must databases of product properties and ways to combine the processes into low-cost separation systems.

One example of a novel and environmentally benign solution was reported by Ladisch and Dyck (1979) in the replacement of benzene or pentane for removing the water in the ethanol-water azeotrope. They used corn as a readily available absorbent for removing the water in the 95 percent ethanol distillate. This approach has been applied only to starch-based ethanol fermentation due to the availability of corn as the raw material. It has been adopted in commercial-scale operations (NRC, 1992). A second novel example, in separation of organic salts produced by fermentation, is the use of electrodialysis combined with bipolar membranes to regenerate the alkaline agent for neutralization and simultaneously produce free acid. This approach avoids creation of undesirable byproduct (Bozell and Landucci, 1993).

New concepts need to be developed and past separation technologies should also be reexamined to reduce costs in downstream processes.

These initiatives should focus on engineering and biological principles as well as combinations of both to improve product purification. New concepts through engineering might use the following approaches:

- supercritical fluid extraction;
- selectively permeable membranes, such as water retentive but solute permeable membranes;
- combination and integration of different separation principles, such as solvent extraction with membrane permeation;
- nonequilibrium or rate-governed processes;
- biomimiary separation technology that capitalizes on the ability of some biological molecules to selectively bind others.

Improved understanding of the inherent abilities of microorganisms—for example, of how solvents and acids impart their toxicity on producing organisms—could provide new ways to reduce downstream processing costs. Yeasts such as *Saccharomyces cerevisiae* can produce ethanol in excess of 150 grams per liter, and bacteria such as *Acetobacter suboxydans* can produce acetic acid (as free acid) in excess of 200 grams per liter. However, the mechanisms and molecular bases for these organisms' ability to tolerate and produce alcohol and acids at these high concentrations is not well understood. Exploration of such biological principles on a more fundamental basis could help improve the commercial viability of fermentation production processes.

SUMMARY

The development of biorefineries will be a key approach to integrating food, feed, chemical, and fuel production in the future. Prototypes of the biorefinery already exist in corn, soybean, and wheat processing plants. These prototype biorefineries process 95 percent of incoming feedstocks into value-added products. Currently 18 to 19 percent of U.S. corn is industrially processed in prototype biorefineries. Pulp mills also represent a biorefinery prototype. Over the near term, stand alone biorefineries will be built around existing wet corn mills. Other near-term opportunities include sugar and wood product manufacturing plants as biorefineries. Future opportunities for biorefineries will be processing plants that efficiently convert plant lignocellulosics to liquid fuels and chemicals based on yet-to-be-proven, high-volume, low-cost processes. Biorefineries offer a number of potential advantages over petroleum refineries, including domestic raw materials, lower environmental impacts, and potentially greater sustainability.

Thermal, mechanical, chemical, and biological processes all play a role in the conversion of biological raw materials into industrial products. Some processes have been proven and used in the past but are not in wide use today. Examples of thermal, mechanical, and chemical processes include liquefaction, destructive distillation to methanol, fast pyrolysis of wood, biomass desiccation, protein fractionation, and chemicals from pentose sugars (e.g., furfural).

Biological processes also show great potential. The fermentation industry currently produces alcohol, organic acids, polymers, antibiotics, enzymes, and amino acids. Advances are occurring in reactor design and the selection and genetic modification of microorganisms. Modern tools of molecular biology are being used to improve yield by increasing product concentration and minimizing undesired byproducts. The major U.S. effort to compete internationally in the amino acid industry provides an example of such advances. Poly(hydroxybutyrate), a biopolymer, is produced by microorganisms and is being introduced into high-value niche markets. Enzymes are being used, for example, in the corn-ethanol industry to convert starch to glucose. Possibly the greatest success to date of enzymatic processing is the use of immobilized glucose isomerase to produce high-fructose corn syrup as a sugar substitute.

The development of lignocellulose treatments will be key to unlocking a major sugar source for biological conversion into industrial products. A number of new potential technologies exist, such as clean fractionation, AFEX, liquid hot water, and perhaps others. For example, the high-fructose corn syrup industry has demonstrated that immobilized enzyme processing and chromatographic separation of biomass can be efficient and economical on a large scale. More research is needed, however, to demonstrate the commercial feasibility of these approaches and to identify new processes.

Other research challenges lie ahead in the development of processing technologies. Advances are needed in the engineering principles for fermentation operations. Biological and engineering approaches should be combined to improve product yield, selectivity, productivity, and product purification. Finally, expansion of our fundamental knowledge of microbial physiology, biochemistry, and genetics will be essential in the development of improved biological processes.

For large-scale biobased products, the dominant factors influencing market share will be the cost of the starting raw material and the cost of processing technology to convert the raw material to the desired biobased products. Further reductions in the price of the biobased raw materials will help process economics but not as much as will reducing the costs of processing technologies. Very low-cost processing technologies must be developed if biobased industrial products are to penetrate commodity markets.

5

Making the Transition to Biobased Products

By the end of the next century, many current petroleum-derived products could be replaced by less expensive and better-performing products based on renewable materials grown in America's forests and agricultural fields. The committee believes that movement to a biobased production system is a sensible approach for achieving economic and environmental sustainability. Biobased products have the potential for being more environmentally friendly because they are produced by less polluting processes than in the petrochemicals industry. Some rural areas may be well positioned to support regional processing facilities dependent on locally grown biobased crops. As a renewable energy source, biomass does not contribute to carbon dioxide in the atmosphere in contrast to fossil fuels. An investment in biobased industries could prepare the nation for a long-term disruption in supplies of imported oil and help to diversify feedstock sources that support the nation's industrial base. These potential benefits of biobased products could justify future public policies that encourage a transition to renewable raw materials for the production of organic chemicals, fuels, and materials.

Despite the potential benefits from biobased products, certain impediments could hamper a transition to biobased production. The carbon-based industries of today are well established and profitable and largely rely on low-priced fossil feedstocks. Introduction of innovative processing technologies has contributed to large returns on investment in petrochemical industries. These energy and chemical companies are vertically integrated to coal, oil, and natural gas and have economic ties to the extraction of these fossil resources. Yet industries constantly transform themselves. For example, witness the Dow/Cargill joint venture to commercialize polylactic acid polymers (biomaterials) derived from corn starch. This venture represents the beginning of an important transformation in feedstocks and processing technologies for the chemical industry. In most cases, however, there is a lack of industrial experience in large-scale processing of complex plant materials. Volatility in petroleum prices continues to be a barrier to the development of these materials. If the government chooses to accelerate development of a biobased industry, well-established petroleum firms may need some incentive to invest in riskier precommercialization stages of development of biobased products.

There may be a compelling national interest to make a transition to a biobased industry. For example, policymakers may want to accelerate use of renewable biomass to mitigate impacts on the U.S. economy from a long-term disruption in world oil supplies or perhaps to reduce impacts on the environment created by possible global warming. However, the degree of public-sector involvement to encourage the growth of a biobased products industry will be a public policy decision. Federal support of research could be a way to make biobased products more competitive. This report makes some recommendations to facilitate research and development (R&D) and commercialization of biobased industrial products.

A VISION FOR THE FUTURE

The committee has described circumstances that it believes will accelerate the introduction of more sustainable approaches to the production of industrial chemicals, liquid fuels, and materials. In this vision a much larger competitively priced biobased products industry will eventually replace much of the petrochemicals industry. The committee proposes intermediate (2020) and long-term (end of century) targets for a future biobased industry. These are summarized below and in Table 5-1:

- by the year 2020, provide at least 25 percent of 1994 levels of organic carbon-based industrial feedstock chemicals and 10 percent of liquid fuels from a biobased products industry;
- eventually satisfy over 90 percent of U.S. organic chemical consumption and up to 50 percent of liquid fuel needs with biobased products; and
- form the basis for U.S. leadership of the global transition to biobased products with accompanying environmental benefits.

Tables 5-2, 5-3, and 5-4 outline the current status of biobased products

TABLE 5-1 Targets for a National Biobased Industry

Biobased Product	Biobased Production Levels (Percent Derived From Biobased Feedstocks)		
	Current Level	Future Target: Intermediate (2020)	Future Target: Ultimate (2090)
Liquid fuels[a]	1–2%	10%	Up to 50%
Organic chemicals[b]	10%	25%	90+%
Materials[c]	90%	95%	99%

[a] Large-scale production of biobased ethanol is a long-term possibility; this projection assumes advanced technologies are in place for processing lignocellulosic materials.

[b] Biobased organic chemicals represent an important market for the biobased industry. Examples include oxygenated chemicals such as butanol or butyl butyrate that can be processed into other intermediate and specialty chemicals traditionally dependent on fossil fuel feedstocks.

[c] Biobased materials includes a wide range of materials extracted directly from plants. Some examples include traditional forest products such as lumber, as well as novel biopolymers such as bioplastics. Many new products in this market will be high-value materials that cannot be produced from petroleum feedstocks.

and some potential actions that could be put into place to meet these targets.

These are intermediate and ultimate targets that are based on estimates of available feedstocks and assume technological advances are in place to improve the suitability of raw materials and conversion processes. While these targets are difficult, they are attainable goals for the following reasons: (1) productivity will almost certainly continue to increase, so more plant material will be produced from less land, (2) food/feed ingredients such as protein will be coproduced with some of the herbaceous energy crops (like alfalfa), and (3) use of agricultural wastes as raw materials for biobased products will reduce competition for resources. It is likely that biobased products will penetrate higher-value chemical markets first. As the technology improves and costs decline, eventually higher-volume, lower-value fuel markets would be penetrated. Demand for biobased materials will continue to grow. Ultimately, the outcomes will be determined by the rate of investment by the private sector.

In the long term, large-scale production of biobased ethanol may supply up to 50 percent of liquid fuel needs in the United States. Once the technology to produce ethanol and other oxygenated chemicals from lignocellulosics becomes economical, the demand for biobased organic chemicals and liquid fuels could increase, creating competition for land and other resources. Coproduction of human food and animal feed prod-

TABLE 5-2 Steps to Achieve Targets of a National Biobased Industry: Biobased Liquid Fuels—Production Milestones

	Current	Future: Intermediate (2020)	Future: Ultimate (2090)
Sources	Corn starch; plant oils and animal fats; wood.	Replace starch and oils with lignocellulosics from wastes and traditional plants and trees (e.g., corn stover, switchgrass, hybrid poplar).	Genetically modified plants and trees provide optimum feedstocks for biorefineries.
Products	Ethanol used in making 10% oxygenated fuels; fatty acid methyl esters comprise 10% of biodiesel; wood for stoves and furnaces.	Ethanol produced at $0.58 per gallon; animal feed coproducts.	Very inexpensive ethanol (less than $0.50 per gallon); animal feed coproducts; many other coproducts.
Processes	Corn wet-milling enzymes; traditional microbes with known processes; transesterification.	Low-cost pretreatment; transgenic microorganisms use C-5 and C-6 sugars.	Combination of physical, chemical, and biological processing in biorefineries minimizes costs.
Status of research and development	Clean Air Act mandates oxygenated fuels; fuel ethanol subsidies; low level of R&D investment by public sector.	Support by government and industry makes lignocellulosics competitive.	Most R&D supported by biobased industry and conducted in academia, government, and industry partnerships; negligible investment by public sector.

TABLE 5-3 Steps to Achieve Targets of a National Biobased Industry: Biobased Organic Chemicals—Production Milestones

	Current	Future: Intermediate (2020)	Future: Ultimate (2090)
Sources	Starch in grains; oils and fats.	Lignocellulosics in wastes and traditional plants and trees; waste sugars.	Genetically modified plants and trees provide optimum sources for biobased products.
Products	Glycerol, ethanol, sorbitol, acetic acid, citric acid, succinic acid, amino acids, poly(hydroxybutyrate), polylactate; detergent enzymes; water-soluble polymers.	Ethanol as feedstock for ethylene; other major oxygenated chemicals; some specialty chemicals such as chiral compounds.	Most specialty, intermediate, and commodity chemicals.
Processes	Corn wet milling; enzymes; traditional microorganisms; thermal and chemical processes.	Low-cost pretreatment of lignocellulosics; transgenic microorganisms; some direct extraction of plant chemicals.	Biorefineries incorporate low-cost, large-scale thermal, chemical, and biological technologies for biomass conversion; multiple chemicals as coproducts of fuel ethanol refining.
Status of research and development	Modest support for R&D by public sector (e.g., USDOE, USDA, NSF, NIH); a few initiatives undertaken by industry.	Research by government and industry provides many new possibilities for exploitation by venture capital; some development assistance by government.	Support of R&D by public sector and industry; R&D conducted by partnerships among academia, government, and industry.

ABBREVIATIONS: USDOE, U.S. Department of Energy; USDA, U.S. Department of Agriculture; NSF, National Science Foundation; NIH, National Institutes of Health.

TABLE 5-4 Steps to Achieve Targets of a National Biobased Industry: Biobased Materials—Production Milestones

	Current	Future: Intermediate (2020)	Future: Long-Term (2090)
Sources	Trees; fiber crops (e.g., cotton and kenaf); strawboard.	Trees; new domesticated crops (e.g., milkweed); soy protein; waste paper.	Transgenic plants and trees produce useful polymers.
Products	Lumber and related building materials; paper and apparel; Tencel fiber for apparel.	Nonwoven fabrics; yarn; paper; composites.	Polymers for plastics; packaging.
Processes	Sawmills; plywood, particle board manufacture; pulp and paper mills; rayon processing.	Harvesting; fractionation; spinning.	Biopolymers; fractionation of transgenic plants.
Status of research and development	Modest investment by public sector (e.g., USDA, DOE) and industry (lumber/paper companies).	Substantial public and industry support of research and development; venture capital funds development of promising processes.	Support of R&D by government and industry; R&D conducted by partnerships among academia, government, and industry.

ucts such as protein with biobased liquid fuels, organic chemicals, and materials is expected to help prevent future conflicts between production of food and biobased products.

Biobased organic chemicals currently constitute 10 percent of the total organic chemicals market. This market may represent the greatest opportunity for replacement of petrochemicals with renewable resources. If ethanol fermentation becomes competitive, enough lignocellulose materials are available to yield ethanol and other oxygenated chemicals (such as butanol or butyl butyrate) that can be further processed into other intermediate and specialty chemicals (Table 5-1). Table 5-3 shows predicted markets for several large-volume chemical and material products. Up to 100 million metric tons of crop residues could be converted into biobased organic chemicals. Other potential resources include production of low-risk crops on a portion of the 35 million acres of land set aside in the Conservation Reserve Program. If one-half of CRP land became available for the production of perennial grasses such as switchgrass, approximately 46 million tons of additional biomass (assuming yields of 2.5 tons per acre) could be available for conversion. The total biomass is sufficient to meet current demands for biobased industrial chemicals.

The biobased materials industry constitutes a major portion of the biobased market. In 1992 wood and paper products accounted for 90 percent of the agricultural and forestry materials used in manufacturing (ERS, 1997b). New products using these traditional materials are under development. The demand for more biobased materials such as bioplastics and novel biopolymers is expected to grow for several reasons. These products are naturally diverse and biodegradable and, compared to biobased commodities (liquid fuels and organic chemicals), specialty chemicals and biopolymers are of higher value and require smaller acreages.

In the long term, development of a strong biobased industry will depend on products that can compete in the marketplace without incentives. A sustained commitment will be required and efforts will need to be integrated among scientists, engineers, economists, raw material producers, processors, manufacturers, marketers, financiers, and business managers. The committee believes that replacement of fossil-based industrial products with renewable materials would be accelerated by public- and private-sector efforts to raise public awareness, focus investment in research and commercialization, and address new approaches to the innovation process.

INVESTMENTS TO ACHIEVE THE VISION

The goal of research, development, and commercialization (RD&C) activities for a biobased industry should be to convert renewable raw materials by appropriate processing into valuable products that sell at prices exceeding the combined input costs.

Reducing the costs of raw materials will continue to be important. With advances in plant molecular biology, cost reductions will occur by genetic engineering of the source plants to make them better suited for processing or direct use. Large cost reductions for biobased products, however, are more likely to occur through development of effective low-cost processing technologies. Such technologies will be physical, thermal, chemical, biological, informational, and combinations thereof. This section discusses RD&C priorities for raw materials, processing, and products, building on concepts and research needs detailed in earlier chapters. A theme throughout this report is that methods, techniques, and technologies developed for a biobased industry must be both effective and economical. Many technically feasible techniques for processing renewable materials have been developed in the laboratory but have little chance of commercial viability. Providing explicit mechanisms for cooperation between laboratory scientists and process engineers would help avoid this problem and help ensure adoption of effective and economical approaches.

The goal of a biobased system is to be sustainable over time. Sustainability can be partially ensured by designing systems capable of processing a variety of raw materials. This will permit greater regional flexibility to make use of the biomass sources most suited to particular locations. Sustainability also will require careful accounting of all material and energy inputs and outputs into the production and processing system, and assurance that healthy soil—the ultimate production resource for biobased products—will be maintained. Economic and environmental sustainability should be the basis of efforts to improve the raw materials, processes, and products of biobased industries.

Niche Products

Niche products are comparatively smaller-scale products that include novel materials such as bioplastics, fatty acids, and other biopolymers. This market deserves special attention because these are high-value products that do not require large acreages of land. Performance is much more important than price, and product differentiation is high among manufacturers. Particularly important niche products are those yielding significant environmental benefits. "Big bang" products, in contrast, are generally large-scale commodity materials for which selling price is the key feature and little or no product differentiation exists. Capitalization needs for commercial-scale operations are significantly lower for niche products in comparison to commodity materials.

Niche products are typically developed by small businesses led by innovative entrepreneurs. Speed in commercialization is crucial, and support consequently must be available when needed and without excessive delays for new funding cycles. Once the process technology, favorable economics, and product characteristics are established, market penetration can begin comparatively rapidly. Innovative entrepreneurs play a pivotal role in spearheading many new commercial developments.

Commodity Products

For commodity products the goal of research and development is to reduce the costs of raw materials and processing because these have a major effect on product cost. The raw materials are heterogeneous, and more than one product is generally produced. In fact, over time the number of products produced from the same raw material tends to increase. Thus, there is no single product; each product is actually a coproduct. Policymakers should focus on public research investments to encourage development of a biobased industry for coproduction of biomass-derived fuels, organic chemicals, and materials. There is potential that coproduction will increase opportunities to create higher-value products from commodity crops.

Public Investments in Research and Development

In the United States, massive public investment in research and development began during World War II and continues to be supported in specific areas. For many years, basic research was regarded primarily as a responsibility of the public sector, while development and commercialization were regarded primarily as responsibilities of the private sector. A large proportion of public funds for research and development were directed toward national security. The federal government assumed special responsibility for ensuring the commercialization of specific identifiable products. For example, the United States adopted this approach for national defense (e.g., high-performance jet aircraft) and for public health (e.g., the polio vaccine). The national interest may be well served by a similar approach for specific biobased products such as biobased ethanol (Lugar and Woolsey, 1999).

Public investment in basic research continues to garner broad support with little controversy surrounding funds for basic research, including fundamental work on process engineering (essential to launch biobased industries). However, public support for development activities that private firms would undertake anyway is not justified (NRC, 1995), and even when private support is uncertain, the use of public funds for development may still be uncertain. It may be less risky and more effective for the public sector to develop the technology to the point where the potential applications are attractive to companies that understand the market. However, a recent report by the U.S. Congress (1998) on national science policy concluded that the need for the government to focus on its critical role in funding basic research is creating a gap between federally funded basic research and industry-funded applied research and development. This gap often is referred to as the Valley of Death. The congressional committee that authored the report concluded that the private sector must recognize and take responsibility for the performance of research but that in some cases of national interest supplementary funds may be justified to provide federal assistance for commercialization of particular technologies. Because of the potential benefits from the expansion of biobased industries, the committee believes limited public involvement in the development of promising technologies is justified.

Between basic research and final commercialization of a new product or a product that directly substitutes for a fossil-based product, the most difficult step is proving the concept at a sufficient scale to encourage full-scale production. This is a key step for attracting the necessary level of private-sector investment for commercialization of emerging technologies. An approach in addressing this concern is through public-private partnerships (Cohen, 1997). One formal mechanism for such partnerships is the Advanced Technology Program (ATP) of the National Institute of Standards and Technology. The ATP is a partnership between government and private industry to accelerate the development of high-risk technologies that promise significant commercial payoffs and widespread benefits for the economy. This committee envisions similar partnerships to facilitate and support biobased research and, in some key cases, government could make an important contribution to proof of concept.

There is limited investment for proof of concept of biobased technologies by the public sector. The ATP helps bridge this gap for selected technologies, but its mission is much broader than biobased technologies. The Alternative Agricultural Research and Commercialization Corporation (AARCC) is a venture capital firm that makes investments in companies to help commercialize biobased industrial products (nonfood, nonfeed) from agricultural and forestry materials and animal byproducts. In its first five years of operation the corporation invested $33 million in federal funds and leveraged $105 million in private funds in 70 projects in 33 states (http://www.usda.gov/aarc/aarcinfo.html).

Subsidizing the development of private-sector products can be controversial. The committee envisions a government-industry partnership in which the government facilitates and supports research and, in key cases where industry will not risk responsibility, government may be a joint supporter at the proof-of-concept stage. These partnerships should emphasize those technologies that are essential to the development of new products and processes across several industries and in cases where the private sector will not risk sole responsibility. R&D efforts should be targeted to those programs and collaborations that can best perform the task. Moreover, these programs should receive periodic evaluation and include endpoint provisions.

An important contribution could be made to the proof-of-concept stage through investment in one or more multifunctional demonstration facilities. Such facilities would house a wide range of flexible large-scale processing equipment and ample qualified support personnel. The facilities might be established by entrepreneurs, industry consortia, or the public sector, with the last possibility most likely. Several places and organizations already exist that could form the nucleus for public-sector-assisted facilities for developing biobased products. Examples include the USDOE Alternative Fuel User Facility in Golden, Colorado; the USDA Laboratory for Agricultural Utilization Research in Peoria, Illinois; and MBI International (formerly the Michigan Biotechnology Institute) in Lansing, Michigan. Regardless of where these facilities might be sited, at federal laboratories or elsewhere, they should be required to obtain a significant fraction of their funds for proof of concept from the private sector. This would maximize the likelihood that commercialization would eventually occur and should bring a degree of market discipline to the process. These facilities could also serve as repositories for the process modeling hardware, software, and databases required to appropriately model new systems or provide sites for analyzing process economics to help set research priorities. These facilities would increase their support of scale-up needs of external clients by providing outside organizations with easier access to facility equipment and technical expertise.

Federal-State Cooperation

Biobased industrial development across the United States often will be region or state specific because of differences in agricultural or forestry resources. Consequently, a diversity of approaches to the development of biobased industries is encouraged. Flexible mechanisms to encourage cooperation between federal and state governments would help achieve this goal.

Incentives

Government agencies (federal, state, or local) can use incentive programs as a mechanism to catalyze biobased industries because the adoption of biobased products requires changing industrial practices and consumer behavior. Such programs might include:

- procurement policies that promote biobased products;
- tax abatement and investment tax credits to accelerate development of biobased industries;
- partial financial support of highly promising biobased technology at the proof-of-concept stage, such as through the AARCC of the USDA (see previous discussion of AARCC activities);
- expansion of investment programs designed to stimulate small businesses in the area of biobased product development; and
- assurance of government research funding that addresses the development of biobased products and processes.

Incentive programs can have widespread implications for the economy, and these effects should be carefully considered by government agencies in developing policies for biobased industrial products. Because the costs of financing some of these incentives are not well known, agencies should incorporate cost-benefit analyses in their decisionmaking. Incentive programs should be cost effective with termination points to evaluate program utility. Ultimately, development of biobased products that can compete in an open market without subsidies is key to sustaining a strong biobased industry.

A government procurement policy is in place to promote biobased products. Under Executive Order 13101, federal agencies and the military have been ordered to implement cost-effective procurement preference programs favoring the purchase of recycled products and environmentally preferable products and services. This mandate encompasses biobased products that utilize biological products or renewable domestic agricultural or forestry materials. Such products may range from building materials and cleaning agents to oils, lubricants, and automotive supplies. The impact of this procurement policy on biobased product development should be evaluated periodically.

Increasing consumer awareness for biobased industrial products could lead to widespread acceptance of biobased products. A seal of authenticity, for example, could raise awareness of environmentally friendly biobased products. National environmental achievement awards could be presented to industries that contribute to the development of technological innovations such as biodegradable plastics that replace polystyrene in single-use disposable materials; biobased solvents that replace toxic chlorinated compounds as degreasing solvents; or bioprocessing routes for nylon production that do not yield nitrous oxide emissions, a cause of acid rain.

Subsidies can distort market forces and have far-reaching effects. Production of biobased chemicals and fuel from lignocellulosics that is cost competitive with gasoline *without subsidies* is a primary goal of R&D for biobased products. The committee believes this to be an achievable goal given the necessary research investment. Should abatement of greenhouse gases become serious, government incentives to develop biobased fuels could be instituted. While the committee recognizes that subsidy decisions are political, not technical, we would argue that, in the long run, subsidies are not a desirable way to support biobased industries or petroleum-derived products.

PROVIDING A SUPPORTIVE INFRASTRUCTURE

To hasten and ease the transition to a biobased economy, a variety of investments should be made in education, technical training, and development of databases. This section examines some of the changes in education, training, and information infrastructure that will be needed to support a biobased industry. Some federal agency might take responsibility for some of this activity or might fund a university consortium to collect and disseminate information on such components as training and curriculum development and biotechnology databases.

Education of the Public

The public as well as policymakers should be educated regarding the rationale and benefits of biobased production. Elected officials and industry leaders in particular must be educated to enable the paradigm shift required for a transition to a biobased economy. Early adoption of this view by private- and public-sector leaders would help generate the required funding support, hasten the transition, and thereby minimize possible dislocations.

One possible way to educate decisionmakers, who develop policy, and the public, which pays for it, would be through an "Annual Conference on Biobased Products Technology." Such a conference could bring together companies, academia, government, and other stakeholders to identify targets and recognize landmark achievements. An important function of this annual meeting and its associated support organizations could be to set standards for biobased materials, perhaps by issuing a seal of authenticity for biobased products.

Technical Training

Today's curricula in chemical and process engineering are thoroughly tied to petroleum processing. Curricula should be revised and then implemented that use examples from and illustrate large-scale processing of renewable materials, including unit operations. Increased curricular flexibility would allow chemical engineers to become better acquainted with the principles and terminology of the biological sciences that are essential to understanding renewable materials. Conversely, the principles and terminology of process engineering also need to be taught to biologists, biochemists, and microbiologists so that engineers and life scientists can better work together to develop the technical infrastructure for developing, manufacturing, and using biobased products.

Improved communication of process engineers with plant breeders and molecular biologists would help tailor various raw materials to specific processing needs. Today innovations for industrial applications are not the focus of most plant breeders because industrial use of plant material is small relative to its use as food and feed. Corn breeders, for example, focus mainly on developing the higher-yield cultivars desired by farmers rather than on modifying corn for industrial applications.

The pool of trained people is scant in certain vital areas, such as natural products chemistry and carbohydrate chemistry. Encouraging an expanded presence for these disciplines on university campuses and industry would speed the development of processes and products requiring such expertise.

Information and Databases

Readily accessible databases could help promote the development of biobased products. Some of the needed information resources are:

- bibliography covering the literature of this field as background to guide future research;
- lists of federal and state grants and funding sources for biobased product research;
- data on ongoing demonstration and precommercialization projects;
- lists of individuals and organizations in the public and private sectors who are active in developing these products and summaries of their facilities and expertise;
- electronic "bulletin boards" for people working in the area;
- statistics on the commercial penetration of biobased products and processes; and
- lists of key organizations promoting the development of these products, such as the New Uses Council Board, the AARCC, and the ATP.

Research Priorities

A national research agenda for a biobased industry should include biological and engineering research that supports the development of economically feasible raw materials, products, and processing technologies.

The development of U.S. biobased industries will require a strong base in science and technology. Advances in agriculture have tended to stress production technologies without a parallel interest in technologies for processing and adding value to agricultural products. For many decades, education and research resources in the fields of chemistry and engineering have focused on petroleum-related opportunities.

Research will be a prominent tool in achieving the committee's vision of making these alternative feedstocks more competitive. The government's research programs should be sensitive to major technical and economic roadblocks that impede the progress of biobased industrial products. Expansion of a biobased industry will require a broad base of knowledge from research in fundamental biological and engineering principles to development of practical technologies that support biobased industries. Where basic research is necessary, innovative ideas should be encouraged through a competitive grants program. For example, investigations of functional genomics are promising new areas of molecular biology and plant genetics that fit this category. In other cases, industrial partnerships with the public sector may be an appropriate mechanism to solve certain engineering problems (e.g., economic feasibility of lignocellulosic conversion processes).

Biological Research

A long-term commitment to fundamental biological research relevant to the needs of a biobased industry should be maintained.

The committee identified three priorities in biological research supporting a biobased industry:

- the genetics of plants and bacteria that lead to an understanding of genes that control plant pathways and cellular processes;
- the physiology and biochemistry of plants and microorganisms directed toward improving bioconversion processes and modification of plant metabolism; and
- protein engineering methods to allow the design of new biocatalysts and novel plant polymers.

Easy Processing and Conversion

Research using new molecular techniques combined with conventional plant breeding will make possible unprecedented modifications of plants to facilitate subsequent processing and conversion to desired products. For example, lignocellulosic fiber crops (including trees) may be developed having a lower lignin content—a change that will make the plants easier to hydrolyze to sugars and hence to ferment to ethanol and other products. Corn grain may be developed that yields modified starches having specific desired properties. Some products may also be encapsulated within plants, increasing the ease with which they are separated from the rest of the raw plant material. Plants also provide a natural source of polymers that may be used as is or from which monomers may be produced and used directly or reassembled into other polymers.

Alter Content of Specific Components

A variety of factors—economic considerations, processing convenience, market demand, and others—may require changes in the content of specific components of a raw material or the final product itself. Desired changes might include an increase or decrease in components, such as lignin, proteins, or oils, or the addition of new polymers. Plant breeders have already succeeded in changing plant oil composition to yield new oils having improved properties for specific uses. Such developments will require an unprecedented degree of research collaboration among process engineers, plant breeders, and molecular biologists.

Processing Advances

In order for biobased products to compete more effectively with petroleum-derived products, the cost of processing raw materials to biobased products must be significantly reduced. Engineering research should focus on developing and improving new and existing processing technologies and on integrating technologies that have the potential to significantly reduce costs.

The committee recommends five key targets for engineering research:

- equipment and methods to harvest (independent of weather conditions) and fractionate lignocellulosic biomass for subsequent conversion processes;
- methods to increase the efficiency and reduce the costs of pretreating lignocellulosic biomass for subsequent conversions to fuel and chemicals, including reducing the costs of the cellulase enzymes (as well as other enzymes such as laccases);

- principles of and processing equipment for handling solid feedstocks;
- fermentation technologies to improve the rate of fermentation, yield, and concentration of biobased products; and
- downstream technologies to further react, separate, and purify products in dilute aqueous streams.

The research priorities for biobased processing are diverse and range from methods for raw material harvest to separation of final biobased products from aqueous media. The committee identified specific areas in processing technologies that should be prioritized in research for the delivery of biobased products to the marketplace. Certain advances could improve the economics of biobased production. Others could unlock the potential of vast new sources of raw materials, make possible whole new product types, or minimize waste production. Ultimately, developing biobased production systems will require modeling to provide an integrated view of sources, processes, and products in the evaluation of technical and economic feasibility.

Separation and Fractionation

Separating raw materials in or near the field might reduce costs and increase the efficiency of subsequent processing steps. Agricultural practice already incorporates in-field fractionation. Corn grain, for example, is separated from stalks and stripped from the ears in the field, and the stalks, cobs, and husks are usually left on the land. The development of lignocellulose conversion technology would make harvesting and transporting the stalks, cobs, and husks necessary. Appropriate harvesting equipment is already available. The harvesting of "wet" crops (e.g., alfalfa) containing industrial enzymes could be improved by equipment that dries the fibrous material and thereby avoids hauling excess water. Developers of biobased products should exploit such opportunities for in-field processing and fractionation to deliver raw materials having the greatest intrinsic value at the lowest possible cost.

Almost all of the renewable carbon materials now available for biobased production are "heterogeneous," that is, made up of more than one component. These components generally will have different uses and different intrinsic values. A fundamental axiom of process engineering is that the components of a heterogeneous raw material become increasingly valuable as they are separated from one another. This applies as much to separating the components of crude oil as to separating copper from ore-bearing rock. Thus, one of the chief research priorities for biobased production is to identify effective and economical methods for separating and fractionating the major components of renewable raw materials (such as carbohydrates, oils, proteins, and lignin).

Pretreatment and fractionation of lignocellulosic biomass together are an especially high priority. The development of effective and economical processes for lignocellulose fractionation can have great economic and social impacts because lignocellulosics—grasses, hays, trees, crop wastes, and the organic fraction of municipal solid waste—can be produced in sufficient volumes to generate biobased replacements for a significant portion of today's petroleum-derived fuels and chemicals. Yet such methods currently are relatively less developed than those for corn grain fractionation (or "refining"). Some lignocellulose fractionation may eventually be accomplished where the raw material is produced (i.e., in the field).

Biological Conversion of Raw Materials

Enzymes

In many cases, enzymes are the best way to process renewable raw materials into biobased products. Moreover, enzymes are themselves an important class of biobased products with industrial markets of hundreds of millions of dollars annually (see also Chapter 3). The ability to use specific enzymes as products and as the catalysts for making other products depends on several factors—enzyme separability, activity, and stability. Each factor merits increased RD&C attention so that the activity and stability of enzymes can be tailored to specific end uses and so that enzymes can be more easily separated from mixtures. Perhaps the most important class of enzymes required to develop a large biobased products industry is the cellulases. The research goals listed above for other enzymes are also appropriate for the cellulases. In addition, particular attention should be paid to means of reducing the costs of the cellulases in integrated processes for biobased products.

Microbial Catalysts

Microbial cells and plants produce enzymes in mixtures containing literally thousands of other components. Effective economical methods are required to better separate industrially useful enzymes from these complex mixtures. Furthermore, enzymes must be recovered as active catalysts, a significant processing challenge because enzymes are inherently only marginally stable. A further difficulty is that the phenolic compounds present in plant extracts often can inactivate enzymes. Genetic engineering of plants should be a viable approach to solving some of these technical problems.

Various approaches, involving science and process engineering, can increase or otherwise manipulate enzyme activity and stability. Research on these approaches deserves increased attention and support, especially work incorporating considerations of effectiveness and economic viability in industrial applications. Many approaches that are effective in a laboratory simply have no reasonable chance of ever being economically viable. In contrast, one particularly attractive approach deserving increased attention is engineering a "purification tag" to assist in the fractionation and purification of recombinant enzymes produced by plants and microbes. Natural and genetically engineered microbes can also act as "biocatalysts" to increase product concentrations, production rates, and yields (or selectivity). These three factors are critical for evaluating the performance of any catalyst. As for enzymes, research is needed to improve methods for separating microbes from mixtures and for tailoring their activity and stability to function in specific catalytic settings. While it is likely that nonbiological catalysts could also be important in generating biobased products, as they have for petroleum-derived products, very little research has examined this possibility. Of particular significance would be the development of nonbiological catalysts that function in aqueous media and are useful in converting oxygen-containing compounds.

Fermentation

The microbes used in the manufacture of biobased products are contained in production vessels referred to generically as fermentors or bioreactors. Today's fermentation bioreactors need to be improved so that they are capable of better heat and mass transfer for viscous fermentation broths (see also Chapter 4). Also required are new methods for monitoring biological processes in fermentors and real-time imaging. Better approaches could reduce fermentor volume and increase volumetric productivity, perhaps by integrating several processes in a single vessel. Finally, new concepts and applications would improve process control for fermentation systems, including improved methods for control of sterility, heat removal, improved selectivity and stability of fermenting organisms, recycling of enzymes, and fermenting organisms in very large fermenters (with capacities up to 3,000 cubic meters).

Concentrating Dilute Aqueous Products

Biobased products that result from reactions in water are present at low concentrations and mixed with other reaction products. Energy-efficient methods are needed to separate and concentrate the desired product. Such methods might include improved membrane-based techniques (such as ultrafiltration and electrodialysis), energy-efficient distillation,

ion exchange, molecular sieves, chromatography, biomimiary separation technology, supercritical fluid extraction, and other methods.

Handling Solids

There is a need for more engineering studies on biomass feedstocks relating to methods and equipment for collection, preprocessing, transport, and storage to minimize total costs delivered to the conversion process. Important scientific problems with significant applications will need to be addressed to improve handling techniques for renewable raw materials, which are usually solids. Solids are more difficult to process, store, and handle than liquids and gases and, consequently, lack similarly well-developed methods. Much work has been done over the past two decades to develop low-cost means of handling agricultural residues and hays. Many biological raw materials are fibrous and create special "bridging" problems, that is, they become intertwined and do not flow smoothly out of storage bins and tanks or along transfer lines. The relatively low-bulk density of some renewable materials also makes them more costly to handle and transport than comparable fossil feedstocks. Finally, difficulties in mixing can hinder the processing of biomass in slurries and can also result in high mass transfer costs.

Reuse of Wastes

The hundreds of millions of tons of organic industrial, agricultural, and municipal wastes generated annually in the United States are disposed of at increasing cost; in fact, these wastes represent a significant potential source of renewable raw materials for biobased products. Current practices in the production, processing, and use of fossil carbon create many environmental problems, including the wastes generated at all stages. Biobased production potentially should cause fewer problems than the petrochemicals industry since biological materials break down in the biosphere. For example, return of mineral-rich effluents from biological processing to the land could avoid waste disposal problems and help maintain long-term soil fertility. Research on wastes should be focused on full use of raw materials, improved biobased processing economics, and reduction of waste materials.

Supporting related R&D and providing incentives to use waste materials would accomplish two important objectives. First, the volume and costs of waste disposal could drop, creating significant environmental benefits. Second, working out the science and technology to use waste materials for biobased products would further the transition to renewable raw materials. It will be important to make a complete energy and environmental audit to demonstrate the potential benefits from reuse of wastes for production of biobased industrial products.

Economic Feasibility

The ultimate goal of all RD&C activities recommended here is to convert renewable raw materials derived from our fields and forests by effective low-cost processing into products that can compete directly with products derived from fossil raw materials. Focused research on agricultural production economics, processing system economics, and integration of food, feed, fuel, and chemical production systems is needed.

Delivering Economic and Effective Biobased Products

This report identifies numerous R&D priorities for raw materials, processing, and products for a biobased industry. A theme throughout is that the methods, techniques, and technologies developed must be both effective and economical. Identifying the conditions in which biobased products can be competitive will be important to the long-term viability of the industry. Economics research will help in identifying potential markets for biobased products and technological developments that can exploit these market opportunities.

Studies of the major industrial product markets should be conducted that include statistical demand analyses, pricing studies, and patterns of import protection. Access to potential markets depends on the extent of the competition and the ability to accommodate changing demands associated with business cycles. The market and general equilibrium effects of the anticipated lignocellulose conversion technology also merit examination.

Meeting the Demand for Food and Industrial Products

Coproducing biobased industrial products with food and feed materials will increase the amount of land effectively available for biobased industrial products and should also improve overall system economics. Recommendations for process technology research to develop these coproduction systems were presented above. Important economic research questions about such coproduction systems might include, for example, the impacts of new protein feeds from biobased industrial products on existing feed markets, land use/availability impacts of coproduction, and overall trends in protein/calorie use/availability due to biobased products.

Agricultural policies could encourage planting of mixed perennial grass feedstocks on erosion-prone farmlands. However, economic research is needed to evaluate the potential competitiveness of processing crops derived from marginal croplands in the absence of government programs that increase the values of these lands.

Modeling Production Systems

Modeling will play a key role in developing the systems for producing biobased industrial products. Modeling provides a way to integrate technical processing with economic considerations in order to assess economic viability. It also helps identify the most costly areas of the overall system for improvement by future research. Software for process modeling already exists, but some of the databases, and perhaps unit processes, needed to apply the software to renewable feedstocks are lacking. In addition, the models require further refinement to include the crop production phase of the overall system, so that work on silviculture, and agriculture, and subsequent processing can be advanced together as an integrated whole.

Environmental Research

Evaluation of the environmental impacts of biobased industries should be a research priority. These evaluations should include environmental and energy audits of the entire product life cycle rather than a single manufacturing step or environmental emission.

Development of a biobased industry may produce widespread environmental benefits, but these implications are not well understood. Production of agricultural and forest feedstocks can have positive, negative, or neutral consequences on wildlife, soil, air, and water quality, but these effects depend on many factors, such as previous use of land and crop management practices. In specific instances, biobased technologies are less polluting, and biobased products are biodegradable. To ensure that biobased products fulfill their promise of environmental sustainability, life-cycle assessments of biobased products should be a research priority.

CONCLUSION

Over the past century, industrial products derived from petroleum—plastics, fuels, lubricants, and building materials—gradually replaced similar products that were once derived from renewable biological materials. Now a transition back to biobased products is taking place, driven by issues related to sustainability of natural resources, human health, environmental quality, economics, and national security. During the next century, the results of innovative collaborations between biological scientists and process engineers are likely to affect industrial development as much as past discoveries in the physical and chemical sciences. This report has identified key opportunities for biobased products and the research and policy priorities that could facilitate the transition. With a vigorous commitment from all parties, the United States will be well positioned to reap the benefits of a strong biobased industry.

References

Abbe, B. 1994. The Expanding Array of Environmentally Improved New Use Products from America's Farms. Paper presented at New Uses Council's 1994 Agriculture Summit on New Uses, June 21-22, Washington, D.C.

Abel, P. P., R. S. Nelson, B. De, N. Hoffmann, S. G. Rogers, R. Fraley, and R. N. Beachy. 1986. Delay of disease development in transgenic plants that express the tobacco mosaic virus coat protection gene. Science 232:738-743.

Ahmed, I. 1993. Industrial Utilization of Agricultural Materials: Energy, Economic, and Environmental Benefits of Bioprocessing. Fourteenth Capital Metals and Materials Forum, Agricultural Commodities: Competing with Traditional Metals and Materials, Oct. 21, Washington, D.C. Washington, D.C.: U.S. Bureau of Mines and the Department of the Treasury.

Ahmed, I., and D. J. Morris. 1994. Replacing Petrochemicals with Biochemicals: A Pollution Prevention Strategy for the Great Lakes. Minneapolis: Institute for Local Self-Reliance.

Ainsworth, S. J. 1994. Soaps and detergents. Chem Eng News (Jan.) 74:34-38.

Allison, R. W., and T. A. Clark. 1994. Effect of enzyme pretreatment on ozone bleaching. Tappi J 77:127-134.

Anderson, E. V. 1993 Brazil's program to use ethanol as transportation fuel loses steam Chem Eng News (Oct.) 71:13-15

API (American Petroleum Institute). 1997. Estimated proved world crude oil reserves. Section II, table 1 in Basic Petroleum Data Book. Washington, D.C.: API.

Avgerinos, G. C., and D. I. C. Wang. 1983. Selective solvent delignification for fermentation enhancement. Biotechnol Bioeng 25:67-83.

Bailey, J. E., and D. F. Ollis. 1986. Biochemical Engineering Fundamentals, 2nd edition. New York: McGraw-Hill.

Bain, R. L. 1993. Electricity from biomass in the United States: Status and future directions. Bioresour Technol 46:86-93.

REFERENCES

Beall, D. S., L. O. Ingram, A. Ben-Bassat, J. B. Doran, D. E. Fowler, R. G. Hall, and B. E. Wood. 1992. Conversion of hydrolysates of corn cobs and hulls into ethanol by recombinant Escherichia coli B containing integrated genes for ethanol production. Biotechnol Lett 14:857-862.

Beall, D. S., K. Ohta, and L. O. Ingram. 1991. Parametric studies of ethanol production from xylose and other sugars by recombinant Escherichia coli. Biotechnol Bioeng 38:296-303.

Bliss, C., and D. O. Blake. 1977. Conversion Processes and Costs. Volume V. Silviculture Biomass Farms. Report No. 7347. McLean, VA: Mitre Corporation, Metrex Division.

Bozell, J. J., and R. Landucci, eds. 1993. Alternative Feedstocks Program: Technical and Economic Assessment, Thermal/Chemical and Bioprocessing Components. Washington, D.C.: U.S. Department of Energy, Office of Industrial Technologies.

Bungay, H. R. 1981. Energy, The Biomass Options. New York: John Wiley & Sons.

Cahoon, E. B., and J. B. Ohlrogge. 1994a. Apparent role of phosphatidylcholine in the metabolism of petroselinic acid in developing umbelliferae endosperm. Plant Physiol 104:845-855.

Cahoon, E. B., and J. B. Ohlrogge. 1994b. Metabolic evidence for the involvement of a delta-4-palmitoyl-acyl carrier protein desaturase in petroselinic acid synthesis in coriander endosperm and transgenic tobacco cells. Plant Physiol 104:827-837.

Carrasco, J. E., M. C. Saiz, A. Navarro, P. Soriano, F. Saez, and J. M. Martinez. 1994. Effects of dilute acid and steam explosion pretreatments on the cellulose structure and kinetics of cellulose fraction hydrolysis by dilute acids in lignocellulosic materials. Appl Biochem Biotechnol 45/46:23-34.

Chum, H. L., and A. J. Power. 1992. Opportunities for cost effective production of biobased materials. Pp. 28-41 in Emerging Technologies for Materials and Chemicals from Biomass, R. M. Roswell, T. P. Schultz, and R. Narayan, eds. Symposium Series 476. Washington D.C.: American Chemical Society.

Claar, P. W., II, T. S. Colvin, and S. J. Marley. 1980. Economic and energy analysis of potential corn-residue harvesting systems. Pp. 273-279 in Agricultural Energy, Paper presented at 1980 American Society of Agricultural Engineers' National Energy Symposium.

Cohen, J. 1997. All-star group prescribes partnerships for R&D woes. Science 275:1410.

Committee on Environment and Natural Resources. 1997. Interagency Assessment of Oxygenated Fuels. Washington, D.C.: Office of Science and Technology Policy.

Dale, B. E. 1983. Biomass refining: Protein and ethanol from alfalfa. Ind Eng Product Research and Development 22:446.

Dale, B. E., and M. J. Moreira. 1982. A freeze explosion technique for increasing cellulose hydrolysis. Pp. 31-43 in Fourth Symposium on Biotechnology in Energy Production and Conservation, vol. 12, C. D. Scott eds. New York: Wiley & Sons.

Daneault, C., C. Leduc, and J. L Valade. 1994. The use of xylanases in kraft pulp bleaching: A review. Tappi J 77:125-131.

Datta, R. 1994. Potential and Implications of Biotechnology for the Food and Agriculture Industry. Briefing paper of the Michigan Biotechnology Institute for the Michigan Department of Agriculture.

Davison, B. H., and C. D. Scott. 1988. Operability and feasibility of ethanol production by immobilized Zymomonas mobilis in a fluidized-bed bioreactor. Appl Biochem Biotechnol 18:19-34.

de la Rosa, L. B., S. T. Reshamwala, V. M. Latimer, B. T. Shawky, B. E. Dale, and E. D. Stuart. 1994. An integrated process for production of ethanol fuel from coastal bermudagrass. Appl Biochem Biotechnol 45/46:483-497.

Delzer, G. C., J. S. Zogorski, T. J. Lopes, and R. L. Bosshart. 1996. Occurrence of the gasoline oxygenate MTBE and BTEX compounds in urban stormwater in the United States, 1991-95. U.S. Geological Survey Water-Resources Investigations Report 96-4145. Reston, Va.: USGS.

Dixon, R. K., S. Brown, R. A. Houghton, A. M. Solomon, M. C. Trexler, and J. Wisniewski. 1994. Carbon pools and flux of global forest ecosystems. Science Weekly (Jan 14) 263:185-190.

Doane, W. M., C. L. Swanson, and G. F. Fanta. 1992. Emerging polymeric materials based on starch. Pp. 197-230 in Emerging Technologies for Materials and Chemicals from Biomass: American Chemical Society Symposium Series 476, R. M. Rowell, T. P. Schultz, and R. Narayan, eds. Washington, D.C.: American Chemical Society.

Dormann, P., M. Frentzen, and J. B. Ohlrogge. 1994. Specificities of the acyl-carrier protein (ACP) thioesterase and glycerol-3-phosphate acyltransferase for octadecenoyl-ACP isomers: Identification of a petroselinoly-ACP thioesterase in umbelliferae. Plant Physiol 104:839-844.

Donaldson, T. L., and O. L. Culberson. 1983. Chemicals from Biomass: An Assessment of the Potential for Production of Chemical Feedstocks from Renewable Resources. ORNL/TM-8432. Oak Ridge, Tenn: Oak Ridge National Laboratory.

EIA (Energy Information Administration). 1997. Alternatives to Traditional Transportation Fuels 1996. DOE/EIA-0585 (96). Washington, D.C: U.S. Department of Energy.

EIA (Energy Information Administration). 1998. Petroleum Marketing Monthly October 1998. *http://www.eia.doe.gov/oil gas/petroleum/pet frame.html*

Elliot, D. C., and P. Walkup. 1977. Bench scale research in thermochemical conversion of biomass to liquids in support of Albany experimental facility. NTIS TID 28415. Richland, WA: U.S. Department of Energy.

Elliott, D. C., D. Beckman, A. V. Bridgwater, J. P. Diebold, S. B. Gevert, and Y. Solantausta. 1991. Developments in direct thermochemical liquefaction of biomass: 1983-1990. Energy Fuel 5:399-410.

Emert, G. H., and R. Katzen. 1980. Gulf's cellulose to ethanol process. Chemtech US 10:610-615.

EPA (Environmental Protection Agency). 1994. EPA Project Summary: Waste Reduction Evaluation of Soy-Based Ink at Sheet-Fed Offset Printer. EPA/600/SR-94/144. Washington, D.C.: EPA.

ERS (Economic Research Service). 1990. Alternative Opportunities in Agriculture: Expanding Output Through Diversification. Report No. 633. Washington, D.C.: U.S. Department of Agriculture.

ERS (Economic Research Service). 1993. Industrial Uses of Agricultural Materials: Situation and Outlook. IUS-1. Washington, D.C.: U.S. Department of Agriculture.

ERS (Economic Research Service). 1996b. Industrial Uses of Agricultural Materials, Situation and Outlook Report. Washington, D.C.: U.S. Department of Agriculture.

ERS (Economic Research Service). 1996c. Special Supplement: Provisions of the 1996 Farm Bill. Washington, D.C.: U.S. Department of Agriculture.

ERS (Economic Research Service). 1997a. Agricultural Resources and Environmental Indicators, 1996-97. Washington, D.C.: U.S. Department of Agriculture, Natural Resources and Environment Division.

ERS (Economic Research Service). 1997b. Industrial Uses of Agricultural Materials. Situation and Outlook Report. Washington, D.C.: U.S. Department of Agriculture, Commercial Agriculture Division.

ERS (Economic Research Service). 1997c. Outyear Projections for U.S. Ag Trade and Food CPI. Washington, D.C.: U.S. Department of Agriculture.

REFERENCES

ERS (Economic Research Service). 1999. Outlook for U.S. Agricultural Trade. Washington, D.C.: U.S. Department of Agriculture.

Falch, E. A. 1991. Industrial Enzymes—developments in production and application. Biotechnol Adv 9:643-658.

Fitchen, J. H., and R. N. Beachy. 1993. Genetically engineered protection against viruses in transgenic plants. Annu Rev Microbiol 47: 739-763.

Fujita, Y., H. Awaji, H. Taneda, M. Matsukura, K. Hata, H. Shimoto, M. Sharyo, M. Abo, and H. Sakaguchi. 1992. Enzymatic pitch control in the papermaking process. Pp. 163-168 in Biotechnology in the Pulp and Paper Industry, M. Kuwahara, and M. Shimada, eds. Tokyo: Uni Publishers.

Gallagher, P. W., and D. L. Johnson. 1995. Some New Ethanol Technology: Cost, Competition and Adaptation Effects in the Petroleum Market. Staff paper #275. Iowa State University, Ames.

Gerhartz, W., ed. 1990. Enzymes in Industry. New York: VCH Publishers.

Gilmour, D. J. 1990. Commercial use of microbe extremophiles. Chem Ind L 9:285-288.

Hacking, A. J. 1986. The American wet milling industry. Pp. 214-221 in Economic Aspects of Biotechnology. New York: Cambridge University Press.

Hain, R., H. J. Relf, E. Krause, R. Langebartels, H. Kindl, B. Vornam, W. Wiese, E. Schmeizer, P. H. Schreier, R. H. Stoer, and K. Stenzel. 1993. Disease resistance results from foreign phytoalexin expression in a novel plant. Nature 361:153-156.

Harsch, J. 1992. New Industrial Uses, New Markets for U.S. Crops: Status of Technology and Commercial Adoption. Washington, D.C.: Cooperative State Research Service.

Hayes, D. J. 1995. Biodiesel: Potential Economic Benefits to Iowa and Iowa Soybean Producers. Ames, Iowa: Center for Agricultural and Rural Development.

Hohenstein, W. G., and L. L. Wright. 1994. Biomass energy production in the United States: An overview. Biomass and Bioenergy 6:161-173.

Hokanson, A. E., and R. Katzen. 1978. Chemicals from wood wastes. Chem Eng Prog 74:67-71.

Holtzapple, M. T., and A. E. Humphrey. 1984. Effect of organosolv pretreatment on enzymatic hydrolysis of poplar. Biotechnol Bioeng 26:670-676.

Holtzapple, M. T., J. H. Jun, G. Ashok, S. L. Patibandla, and B. E. Dale. 1991. The ammonia freeze explosion (AFEX) process: A practical lignocellulose pretreatment. Appl Biochem Biotechnol 28/29:59-74.

Horsch, R. B., R. T. Fraley, S. G. Rogers, P. R. Sanders, A. Lloyd, and N. Hoffmann. 1984. Inheritance of functional foreign genes in plants. Science 223:496-498.

Horsch, R. B., J. E. Fry, N. L. Hoffmann, D. Eichholtz, S. G. Rogers, and R. T. Fraley. 1985. A simple and general method for transferring genes into plants. Science 227:1229-1231.

Hunt, J. F., and C. T. Scott. 1988. Combined board properties of FPL spaceboard formed by a new method. Tappi J 71(11):137-141.

Ince, P. J. 1996. Recycling of Wood and Paper Products in the United States. Madison, Wisc.: U.S. Department of Agriculture, Forest Servcie.

Ingram, L. O., and T. Conway. 1988. Expression of different levels of ethanologenic enzymes from Zymomonas mobilis in recombinant strains of Escherichia coli. Appl Environ Microb 54:397-404.

Ingram, L. O., T. Conway, D. P. Clark, G. W. Sewell, and J. F. Preston. 1987. Genetic engineering of ethanol production in Escherichia coli. Appl Environ Microb 53:2420-2425.

Jeffries, T. W., J. H. Klungness, M. S. Sykes, and K. R. Rutledge-Cropsey. 1994. Comparison of enzyme-enhanced with conventional deinking of xerographic and laser-printed paper. Tappi J 77:173-179.

Jenes, B., H. Moore, J. Cao, W. Zhang, and R. Wu. 1993. Techniques for gene transfer. Pp. 125-146 in Transgenic Plants: Engineering, and Utilization, vol. 1, S. Kung and R. Wu, eds. New York: Academic Press.

Kane, S., J. Reilly, M. LeBlanc, and J. Hrubovcak. 1989. Ethanol's role: An economic assessment. Agribusiness 5:505-522.

Kaplan, D. L., J. My Maey, D. Ball, J. McCassie, A. L. Allen, and P. Stenhouse. 1992. Fundamentals of biodegradable polymers. Paper Presented at the Biodegradable Materials and Packaging Symposium, June, Natick, Mass.

Katzen, R., and D. E. Fowler. 1994. Ethanol from lignocellulosic wastes with utilization of recombinant bacteria. Appl Biochem Biotechnol 45/46:697-707.

Kerr, R. 1998. The next oil crisis looms large—and perhaps close. Science 281:1128-1131.

Kirk, T. K., J. W. Koning, R. R. Burgess, M. Akhtar, and R. A. Blanchette. 1993. Biopulping: A Glimpse of the Future. Forest Service Research Paper, FPLRP523. Madison, Wisc.: University of Wisconsin (Madison) and University of Minnesota (St.Paul).

Kishore, G. M., and C. R. Somerville. 1993. Genetic engineering of commercially useful biosynthetic pathways in transgenic plants. Curr Opin Biotechnol 4:152-158.

Kishore, G. M., and D. M. Shah. 1988. Amino acid biosynthesis inhibitors as herbicides. Annu Rev Biochem 57:627-663.

Kleiman, R., and G. F. Spencer. 1982. Search for new industrial oils. XVI. Umbelliflorae—seed oils rich in petroselinic-acid. J Am Oil Chem Soc 59:29-38.

Klibanov, A. M. 1983. Immobilized enzymes and cells as practical catalysts. Science 219:722-727.

Klibanov, A. M. 1990. Asymmetric transformations catalyzed by enzymes in organic solvents. Acc Chem Res 23:114-120.

Knowles, J. R. 1987. Tinkering with enzymes: What are we learning? Science 236:1252-1258.

Ladisch, M. R., and K. Dyck. 1979. Dehydration of ethanol: New approach gives positive energy balance. Science 205:898-900.

Ladisch, M. R., C. M. Ladisch, and G. T. Tsao. 1978. Cellulose to sugars: New path gives quantitative yield. Science 201:743-745.

Layman, P. 1990. Promising new markets emerging for commercial enzymes. Chem Eng News (Sep.) 68:17-18.

Leung, P., Y. M. Lee, E. Greenberg, K. Esch, S. Boylan, and J. Preiss. 1986. Cloning and expression of the Escherichia coli GLgC gene from a mutant containing an ADPglucose pyrophosphorylase with altered allosteric properties. J Bacteriol 167:82-88.

Lipinsky, E. S. 1981. Chemicals from biomass: Petrochemical substitution options. Science 212:1465-1471.

Lugar, R. G. and R. J. Woolsey. 1999. The new petroleum. Foreign Affairs 78:88-102.

Lynd, L. R. 1996. Overview and evaluation of fuel ethanol from cellulosic biomass: Technology, economics, the environment and policy. Annu Rev Energy Environ 21:403-465.

Macray, A. R. 1985. Interesterification of fats and oils. Pp. 195-208 in Biocatalysis in Organic Solvents, J. Tramper, H. C. van der Plas, and P. Lindo, eds.

Maiorella, B., H. W. Blanche, and C. R. Wilke. 1984. Economic evaluation of alternative ethanol fermentation processes. Biotechnol Bioeng 26:1003-1025.

Maiorella, B., C. R. Wilke, and H. W. Blanch. 1981. Alcohol production and recovery. Adv Biochem Eng 20:43-92.

Marland, G., and B. Schlamadinger. 1995. Biomass fuels and forest management strategies: How do we calculate the greenhouse-gas emission benefits? Energy 20:1131-1140.

McBride, K. E., D. J. Schaat, M. Daley, and D. M. Stalker. 1993. Fourth International Congress of Plant Molecular Biology. Number 343.

REFERENCES

McKillip, W. J., and E. Sherman. 1979. Furan derivatives. Pp. 499-527 in Kirk Othmer Encyclopedia of Chemical Technology, 3rd ed., vol. 11. New York: John Wiley & Sons.

McMillan, J. D. 1994. Pretreatment of lignocellulosic biomass. Pp. 292-324 in Enzymatic Conversion of Biomass for Fuels Production, M. E. Himmel, J. O. Baker, R. P. Overend, eds. ACS Symposium Series No. 566. Washington, D.C.: American Chemical Society.

Meier zu Beerentrup, H., and G. Roebbelen. 1987. Calendula and coriandrum: New potential oil crops for industrial uses. Fett Wissenschaft Technologie 89:227-230.

Miller, R., B. Heine, C. Jewitt, C. Reeder, and W. R. Schwanst. 1996. Changes in Gasoline III: The Auto Technician's Gasoline Quality Guide. Brenin, Ind.: Bownstream Alternatives, Inc.

Miller Freeman, Inc. 1995. Lockwood-Post's Directory of Pulp, Paper, and Allied Trades. San Francisco: Miller Freeman, Inc.

Mohagheghi, A., K. Evans, M. Finkelstein, and M. Zhang. 1998. Cofermentation of glucose, xylose, and arabinose by mixed cultures of two genetically engineered Zymomonas mobilis strains. Appl Biochem Biotechnol 70/72:285-299.

Morris, D. J., and I. Ahmed. 1992. The Carbohydrate Economy: Making Chemicals and Inidustrial Marterials from Plant Matter. Washington D.C.: Institute of Local Self Reliance.

Narayan, R. 1994. Impact of governmental policies, regulations, and standards activities on an emerging biodegradable plastics industry. In Biodegradable Plastics and Polymers, Y. Doi, and K. Fukuda, eds. New York: Elsevier.

NASS (National Agricultural Statistics Service). 1994. Field Crops: Final Estimates by States, 1987-92. U.S. Department of Agriculture, Agricultural Statistics Board, Statistical Bulletin No. 896. Available online: *http://usda.mannlib.cornell.edu/datasets/crops/94896/sb896.txt*

NRC (National Research Council). 1989. Field Testing Genetically Modified Organisms. Washington, D. C.: National Academy Press.

NRC (National Research Council). 1992. Putting Biotechnology to Work: Bioprocess Engineering. Washington, D.C.: National Academy Press.

NRC (National Research Council). 1995. Allocating Federal Funds for Science and Technology. Washington, D.C.: National Academy Press.

NRC (National Research Council). 1996a. Ecologically Based Pest Management. Washington, D.C.: National Academy Press.

NRC (National Research Council). 1996b. Toxicological and Performance Aspects of Oxygenated Motor Vehicle Fuels. Washington, D.C.: National Academy Press.

NRC (National Research Council). 1997. Wood in Our Future: The Role of Life-Cycle Analysis. Proceedings of a Symposium. Washington, D.C.: National Academy Press.

Office of Science and Technology Policy. 1997. Interagency Assessment of Oxygenated Fuels. Washington, D.C.: Committee on Environment and Natural Resources.

Ohlrogge, J. B. 1994. Design of new plant products: Engineering of fatty acid metabolism. Plant Physiol 104:821-826.

O'Neil, D. J. 1978. Design Fabrication and Operation of a Biomass Fermentation Facility. First quarterly report to the U.S. Department of Energy. Atlanta; Georgia Institute of Technology, Engineering Experiment Station.

Osborn, T. 1997. New CRP criteria enhance environmental gains. Agricultural Outlook (Oct.) 245:15-18.

OTA (Office of Technology Assessment). 1980. Energy from Biological Processes. Vol. II: Technical and Environmental Analyses. Washington, D.C.: U.S. Government Printing Office.

OTA (Office of Technology Assessment). 1993. Potential Environmental Impacts of Bioenergy Crop Production—Background Paper. OTA-BP-E-118. Washington, D.C.: U.S. Government Printing Office.

Park, Y. W. 1997. Economic Feasibility of Growing Herbaceous Biomass Crops in Iowa. Ph.D. dissertation, Iowa State University, Ames.

Perlak, F. J., R. W. Deaton, T. A. Armstrong, R. L. Fuchs, S. R. Sims, J. T. Greenplate, and D. A. Fischhoff. 1990. Insect resistant cotton plants. Bio-Technol 8:939-943.

Pirie, N. W., ed. 1971. Leaf Protein: Its Agronomy, Preparation. Quality and Use. IBP Handbook No. 20. London: International Biological Programme.

Poirier, Y., C. Nawrath, and C. Somerville. 1995. Production of polyhydroxyalkanoates, a family of biodegradable plastics and elastomers, in bacteria and plants. Bio-Technol 13:142-150.

Polman, K. 1994. Review and analysis of renewable feed stocks for the production of commodity chemicals. Appl Biochem Biotechnol 45/46:709-722.

Roberts, D. 1998. Preliminary Assessment of the Effects of the WTO Agreement on Sanitary and Phytosanitary Trade Regulations. Journal of International Economic Law 1:377-405

Rooney, T. 1998. Lignocellulosic Feedstock Resource Assessment. Report SR-580-24189. Golden, Colo.: National Renewable Energy Laboratory.

Schultz, P. G., R. A. Lerner, and S. J. Benkovic 1990. Catalytic antibodies; Special Report. Chem Eng News (May) 68:26-40.

Scott, C. D., B. H. Davison, T. C. Scott, J. Woodward, C. Dees, and D. S. Rothrock. 1994. An advanced bioprocessing concept for the conversion of waste paper to ethanol. Appl Biochem and Biotechnol 45/46:641-653.

Smith, W. B., J. L. Faulkner, and D. S. Powell. 1994. Forest Statistics of the United States, 1992 Metric Units. Forest Service General Technical Report NC-168. Washington, D.C.: U.S. Department of Agriculture, North Central Forest Experiment Station.

Squillace, P. J., J. S. Zogorski, W. G, Wilber, and C. V. Price. 1996. Preliminary assessment of the occurrence and possible sources of MTBE in groundwater in the United States, 1993-1994. Environ Sci Technol 30:1721-1730.

Stamm, A. J. 1964. Wood and Cellulose Science. New York: Ronald Press.

Stark, D. M., K. P. Timmerman, G. F. Barry, J. Preiss, and G. M. Kishore. 1992. Regulation of the amount of starch in plant tissues by ADP glucose pyrophosphorylase. Science 258:287-292.

Staskawicz, B. J., F. M. Ausubel, B. J. Baker, J. G. Ellis, and J. D. G. Jones. 1995. Molecular genetics of plant disease resistance. Science Weekly (May) 268:661-667.

Stinson, S. C. 1979. Methanol primed for future energy role. Chem Eng News (April) 57:28-30.

Szmant, H. H. 1987. Industrial Utilization of Renewable Resources. Lancaster, Pa.: Technomic Publishing

Tarczynsky, M. C., R. G. Jensen, and H. J. Bohnert. 1993. Stress protection of transgenic tobacco by production of the osmolyte mannitol. Science 259:508-510.

Terras, F. R. G., H. M. E. Schoofs, M. F. C. De Bolle, F. van Leuven, S. B. Rees, J. Vanderleyden, B. P. A. Cammue, and W. F. Broekaert. 1992. Analysis of two novel classes of plant antifungal proteins from radish (Raphanus sativus L.) seeds. J Biol Chem 267: 15301-15309.

Thayer, A. 1994. Novo Nordisk opens major enzyme plant. Chem Eng News (April) 72:9.

Tirrell, D. 1996. Putting a new spin on spider silk. Science 271:39-40.

Tolbert, V. R., and A. Schiller. 1996. Environmental enhancement using short-rotation woody crops and perennial grasses as alternative agricultural crops. Conference on Policy and Marketing: Positioning Ethanol, ETBE and E-85 for the 21st Century. CONF-9511191-1. Washington. D.C.: U.S. Department of Energy.

REFERENCES

Tolbert, V. R., J. E. Lindberg, T. H. Green, R. Malik, W. E. Bandaranayake, J. D. Joslin, F. C. Thornton, D. D. Tyler, A. E. Houston, D. Petiry, S. Schoenholtz, B.R. Bock, and C. C. Trettin. 1997. Proceedings of the International Workshop on Environmental Aspects of Energy Crop Production. Bra Simone, Italy, October 9-10, 1997.

Topfer, R., N. Martini, and J. Schell. 1995. Modification of plant lipid synthesis. Science 268:681-686.

Tsai, F. Y., and G. M. Coruzzi. 1993. Transgenic plants for studying genes encoding amino acid biosynthetic enzymes. Pp. 181-194 in Transgenic Plants, Engineering, and Utilization, vol. 1, S. Kung and R. Wu, eds. San Diego: Academic Press.

U.S. Congress. 1998. Unlocking Our Future: Toward a New National Science Policy. Report to Congress by the House Committee on Science. *http://www.house.gov/science/science policy report.htm.*

USDA (U.S. Department of Agriculture). 1992. New crops, new uses, new markets: Industrial and commercial products from U.S. agriculture. Yearbook of Agriculture series. Washington, D.C.: USDA.

USDA (U.S. Department of Agriculture). 1993. Agricultural Statistics. Washington, D.C.: U.S. Government Printing Office.

USDA (U.S. Department of Agriculture). 1994. Ethanol and citric acid increase the use of corn. Pp. 14-16 in Industrial Uses of Agricultural Materials: Situation and Outlook Report. Publication IUS-3. Washington, D.C.: USDA.

USDA (U.S. Department of Agriculture). 1995. The 1993 RPA Timber Assessment Update. General Technical Report RMGTR-259. Ft. Collins, Colo.: USDA.

USDA (U.S. Department of Agriculture). 1997a. Agricultural Baseline Projections to 2005, Reflecting the 1996 Farm Act. Prepared by the Interagency Agricultural Projections Committee. Staff Report No. WAOB-97-1. Washington, D.C.: USDA.

USDA (U.S. Department of Agriculture). 1997b. USDA's 1997 Baseline: The Domestic Outlook to 2005. Agricultural Outlook. Economic Research Service. April 1997. AO-239. Washington, D.C.: USDA.

USDOE (U.S. Department of Energy). 1998. Comprehensive National Energy Strategy. *http://www.hr.doe.gov/nesp/cnes.htm.*

Van Walsum, P., S. G. Allen, M. J. Spencer, M. S. Laser, M. J. Antal, and L. R. Lynd. 1996. Conversion of lignocellulosics pretreated with hot compressed liquid water to ethanol. Appl Biochem Biotechnol 57/58:157-170.

Wall, J. S., and J. W. Paulis. 1978. Corn and sorghum grain proteins. Pp. 135-219 in Advances in Cereal Science and Technology, vol. 2, Y. Pomeranz, ed. St. Paul, Minn.: American Association of Cereal Chemists.

Walsh, M. E., D. Becker, and R. L. Graham. 1996. The Conservation Reserve Program as a means to subsidize bioenergy crop prices. Pp. 75-81 in Bioenergy '96: Partnership to Develop and Apply Biomass Technologies. Muscle Shoals, Alabama: Tennessee Valley Authority Southeastern Regional Biomass Energy Program.

Webb, O. F., B. H. Davison, T. C. Scott, and C. D. Scott. 1995. Design and demonstration of an immobilized-cell fluidized bed reactor for the efficient production of ethanol Appl Biochem Biotechnol 51/52: 559-568.

Wescott, C. R., and A. M. Klibanov. 1994. The solvent dependence of enzyme specificity. Biochim Biophys Acta 1206:1-9.

White, D. H., and D. Wolf. 1988. Research in Thermochemical Biomass Conversion, A. V. Bridgwater and J. L. Kuester, eds. New York: Elsevier Applied Science

Withers, S. G., Q. Wang, L. Mackenzie, and R. A. J. Warren. 1996. Glycosynthases: Engineered glycosidases which synthesize but do not degrade oligosaccharides. Abstr Pap Am Chem Soc 212:BIOL 090.

Wood, B. E., and L. O. Ingram. 1992. Ethanol production from cellobiose, amorphous cellulose, and crystalline cellulose by recombinant Klebsiella oxytoca containing chromosomally integrated Zymomonas mobilis genes for ethanol production and plasmids expressing thermostable cellulase genes from Clostridium thermocellum. Appl Environ Microb 58:2103-2110.

Wrotonowski, C. 1997. Unexpected niche applications for industrial enzymes drives market growth. Genetic Engineering News 17:35.

Yoshida, Y., T. Kiyosue, T. Katagiri, H. Ueda, T. Mizoguchi, K. Yamaguchi, Shinozaki, K. Wada, Y. Harada, and K. Shinozaki. 1995. Correlation between the induction of a gene for delta 1-pyrroline-5-carboxylate synthetase and the accumulation of proline in *Arabidopsis thaliana* under osmotic stress. Plant J 7:751-760.

Zanin, G. M., L. M. Kamabara, L. P. V. Calsavara, and F. F. Demoraes. 1994. Performance of fixed and fluidized-bed reactors with immobilized enzyme. Appl Biochem Biotechnol 45/46:627-640.

Zeikus, J. G. 1990. Accomplishments in Microbial Biotechnology in Biotechnology Science, Education and Commercialization. New York: Elsevier.

Zhang, M., C. Eddy, K. Deanda, M. Finkelstein, and S. Picataggio. 1995. Metabolic engineering of a pentose metabolism pathway in ethanologenic Zymomonas mobilis. Science 267:240-243.

APPENDIXES

A

Case Study of Lignocellulose Ethanol Processing

The U.S. Department of Energy (DOE) examined the economics of producing ethanol from wood chips in a 1993 report (Bozell and Landucci, 1993). The analysis below is based on that study but considers corn stover—corn stalks, leaves, and husks—as the potential feedstock. Examination of the feedstock supply and demand, and the costs of transportation and processing, suggests that 7.5 billion gallons of ethanol could be produced annually at a cost of about $0.58 per gallon. When corrected for fuel efficiency, the cost to replace a gallon of gasoline becomes roughly $0.58, potentially making ethanol cost competitive without subsidies. This projection critically hinges on new technologies and on low corn residue costs because these residues are coproduced with corn grain. An additional 4.5 billion gallons of ethanol may be produced at potentially higher costs due to higher prices for corn stover.

FEEDSTOCK SUPPLY AND DEMAND

Figure A-1 shows a supply curve for corn stover. Supply curves identify the amount of a resource available in the market at a given price as determined by its value in the best alternative use. The supply price for a new crop typically consists of production costs (planting, care, harvesting) plus an allowance for land rent, where land rent represents the value of the land when it is used to produce another crop. The cost of corn stover may be lower than new crops, at least for some levels of use; stover and corn grain are produced together, making recovery of land

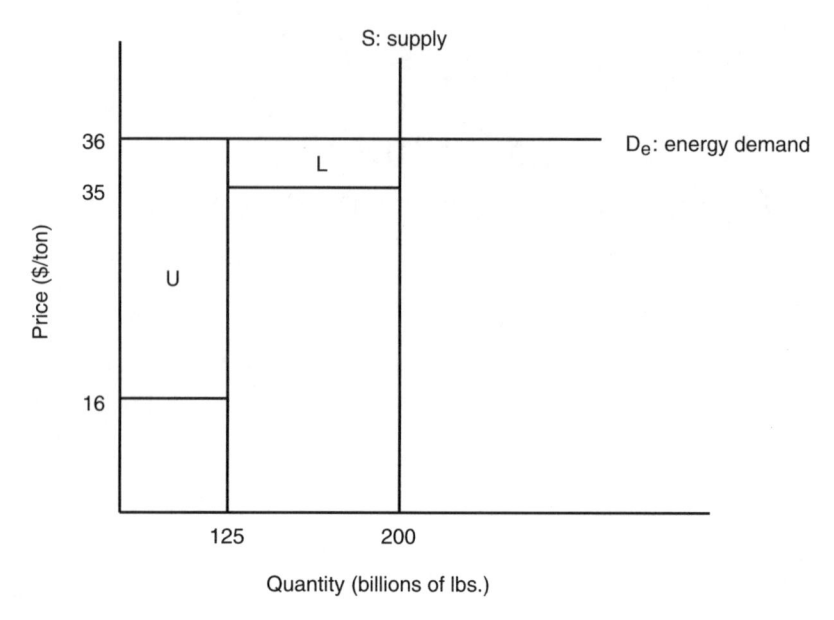

FIGURE A-1   Corn stover supply and demand curve.

costs for corn stover unnecessary because they have already been accounted for in grain profit calculations.

The present economic value of corn stover arises from two sources. First, erosion and fertilizer requirements are reduced when corn stover is left on the ground. Second, corn stover replaces low-grade hay when fed to cows. A first approximation of the stover supply curve facing ethanol processors is a step function (Figure A-1). It begins horizontal at net harvest cost (U). Processing plants in well-chosen locations in cash-grain areas could acquire stover that is not used by livestock producers at slightly above harvest cost. The second step of the curve is the higher value that livestock producers are willing to pay for stover used as feed (L). If processors are willing to pay slightly more than the livestock value, all stover supplies would be diverted to industrial uses. The supply curve is vertical where industry uses all available supplies, provided that the amount of land that is planted with corn is kept fixed.

If the new cellulose conversion technology develops successfully, energy products will create a new market for corn stover. The horizontal demand curve, D_e in Figure A-1, shows energy producers' returns for processing a unit of corn stover (the difference between energy product price and stover processing costs). The market determines corn stover use up to where processors' returns equal corn stover harvest costs—the intersection of supply and demand. If energy prices rise, the demand curve will shift upward, and progressively more corn stover (and other raw materials) will be available at progressively higher prices in a market economy. Estimates of corn stover use opportunity costs to determine the dollar per ton (height) steps of the corn supply curve.

Table A-1 gives net harvest cost estimates based on harvest expenses and fertilizer replacement costs for the midwestern United States. Harvest costs are estimated from hay harvesting costs (fixed machinery replacement costs and variable operating costs). The calculations use a harvested stover tonnage that includes an amount (30 percent) left on the field for soil conservation compliance. The fertilizer estimate is based on replacing phosphorus and potassium contained in the harvested corn stover. The stover net harvest cost is $16.50 per ton. Reported production costs for several energy feedstocks in a Midwest location are considerably higher than harvest cost for corn stover. The estimates range from $42.00 per ton for sweet sorghum to $95.00 per ton for canary grass in Ames, Iowa, where crop rents are relatively high. Costs were lower at Chariton, Iowa, due to reduced land costs, ranging from $36.00 for sweet sorghum to $93.00 per ton for alfalfa. The feed value of hay can be similarly calculated using some adjustments for total digestible nutrients and the protein deficiency of corn stover in comparison to hay. The value of corn stover as a feed is about $35.00 per ton based on the 1994 hay price.

TABLE A-1 Costs of Corn Stover Harvest in the United States, 1993

DIRECT HARVEST COSTS

Operation	Reported Fixed Cost ($)	Fixed Cost ($/ton)	Reported Variable Cost ($)	Variable Cost ($/ton)	Total ($/ton)
Rake	2.43/acre	1.22	1.52/acre	0.76	
Baler	3.14/bale	6.28	2.05/bale	4.10	
Total		7.50		4.86	12.36

INDIRECT FERTILIZER REPLACEMENT COSTS

Fertilizer	Application Rate (lbs./acre)	Price ($/ton)	Total ($/ton)
P_2O_5	13	150	0.49
K_2O	71	206	3.66
Total			4.14

TOTAL: Direct + Indirect Costs = $16.50
SOURCE: Claar et al. (1980).

The volume of corn stover available to the processing industry can be approximated using estimates of available corn stover, cattle population and forage requirements, and the availability of hay for forage. Calculations (not shown) involved multiplying a state's corn area by a stover yield estimate that leaves an allowance for compliance with the Conservation Reserve Program. Similarly, the estimate of cattle feed demand is the product of cattle population and forage requirement per animal, less hay supply for each state. The industry supply is the supply less feed demand—in this case about 125 billion pounds (62.5 million tons) available at low prices near the harvest cost of $16.00 per ton. At prices above the feed cost of $35.00 per ton, the entire stover supply of about 200 billion pounds (100 million tons) would be available to the processing industry.

Approximately 7.5 billion gallons of ethanol could be produced at about $0.46 per gallon from corn stover. An additional 4.5 billion gallons of ethanol may be produced at higher costs due to a potentially higher price for corn stover above the feed cost of $35.00 per ton. The calculations below assume maximum theoretical yields for conversion of pentose and glucose sugars:

$$125 \text{ billion pounds of stover} \times \frac{0.40 \text{ pounds of ethanol}}{1 \text{ pound of stover}} \times \frac{1 \text{ gallon of ethanol}}{6.6 \text{ pounds of ethanol}} = 7.5 \text{ billion gallons}$$

$$200 \text{ billion pounds of stover} \times \frac{0.40 \text{ pounds of ethanol}}{1 \text{ pound of stover}} \times \frac{1 \text{ gallon of ethanol}}{6.6 \text{ pounds of ethanol}} = 12 \text{ billion gallons}$$

When correction is made for the relative fuel efficiency of ethanol and gasoline, the 12 billion gallons of ethanol from corn stover translates to the equivalent of 9.6 billion gallons of gasoline or about 9 percent of annual U.S. gasoline consumption (about 110 billion gallons).

TRANSPORTATION COSTS

A large processing plant could exploit economies of scale but would also require vast amounts of corn stover (about 2.9 million tons for a 350 million gallons per year ethanol plant). Given the average corn density in Iowa, for example, all the corn stover available in a 50 mile radius would be required. Since the delivered cost of the stover increases with distance, the transportation component of input cost could become large and offset economies of scale when such a bulky material is used as a feedstock. The firm's average input costs (AICs) can be approximated by the formula AICs = P_0 + 2tr/3, where P_0 is the harvest cost, r is the radial distance from the plant (in miles), and t is the transportation cost (in dollars per ton per mile). Based on quotations from Iowa trucking firms, the 1994 transportation rate is between $0.10 and $0.15 per ton-mile (for shorter distances). Hence, the average total cost of all stover drawn from within 50 miles is $20.00 to $21.50 per ton.

The lower stover estimate was used to calculate processing cost (below) because the routine of a large plant could reduce the short-haul rate and because the average corn stover density may understate availability at the local level. Some offsetting bias may exist in these calculations; distance is measured "as the crow flies" instead of on a particular road network and therefore may be understated. Also, the average corn density may understate availability at the substate level because it combines low-density crop production from cattle grazing areas of the south with high-density corn production in cash-grain areas of the north. Well-chosen plant locations would be in the cash-grain areas, where stover could be acquired at near harvest costs. Overall, the calculations do suggest that transportation costs are not a major barrier to operation of a large-scale plant, at least in the Corn Belt.

PROCESSING COSTS

Revised material flows and cost estimates are shown in Table A-2. This analysis adjusts reported cost data on bioprocessing for uniformity and for comparisons that indicate tradeoffs by decisionmakers. The cost data also conform to the procedures of Donaldson and Culberson (1983) that facilitate comparison to petrochemical processes. Variable costs include materials and utilities. Labor costs are included where available. Fixed costs are limited to capital costs, calculated as the annual payment on a 15-year, 10 percent interest, fixed annual payment mortgage on the entire plant cost. Other expenditures are excluded from fixed costs, including overhead (because there is no opportunity cost) and insurance (in order to offset risks of the profit stream, which should be considered elsewhere). Expenses such as labor for plant maintenance or taxes on the plant could be included, but individual situations vary, and data were not uniformly available. Cost data for some petrochemical processes (styrene, ethylene, ethanol) were developed using Donaldson and Culberson's estimates of input requirements, yields, and plant costs—combining input requirements with recent price data to estimate material and utility expenditures, updating capital expenditure data with a price index for plant and equipment, and giving annual payment for a 15-year mortgage. These calculations assume that stover is available at near its harvest cost of $16 per ton and includes a transportation cost. It is also assumed that technology for fermentation of pentose sugars is fully implemented.

Other cost adjustments were required for conversion to corn stover

TABLE A-2 Production Cost Estimate for Plant Processing Corn Stover to Ethanol[a]

	Units per 10³ Gallons of Ethanol	1993 Price ($)	Unit Expenses per 10³ Gallons of Ethanol ($)
Wood/stover, ton[b]	8.3	19.805/ton	164.38
Sulfuric acid, pounds	297.0	86.20/ton	12.80
Lime, pounds	219.0	40/ton	4.38
Ammonia, pounds	470.4	200/ton	47.04
Nutrients, pounds	18.1	230/ton	2.08
Corn liquor, pounds	63.3	266/ton	8.42
Corn oil, pounds	3.9	413.4/ton	0.81
Glucose, pounds	37.0	258/ton	4.77
Catalyst	1000.0	0.01/unit	10.00
Disposal, ton	0.34	20/ton	(6.80)[c]
Water, million gallons	19.87	0.002/gallon	0.04
Total Materials			**247.92**
Labor, man-year	41.0[d]	29,800/year	3.49
Foremen, man-year	9.0[d]	34,000/year	0.88
Supervisors, man-year	1.0[d]	40,000/year	0.11
Total Materials + Labor			**252.40**
Capital Allowance[e]			211.4
Total Production Cost			**463.8**
Total Processing Cost (net of stover)			**299.4**

Ethanol output	350 million gallons
Input requirement	2.90 million tons
Plant cost	$562 million

[a] Cost estimates will vary with each operation and computer model used for analysis. This estimate assumes that advanced technology for fermentation of pentase sugars will be fully developed and implemented.

[b] Stover cost is based on harvest cost, including fertilizer replacement, of $16.50 per ton, plus transportation costs of $3.30 per ton.

[c] A byproduct credit (negative disposal cost).

[d] Man-years required for output of 350 million gallons of ethanol.

[e] Capital allowances assume 10 percent return and 15-year amortization; allowances vary with lender.

from wood chips. In particular, the lignin content of corn stover is only one-half of that for wood chips, so the one-half of electrical plant capacity that was sold as a byproduct credit is removed. Also, the 10 percent gasoline mixing operation was eliminated, so estimates now refer to a pure ethanol basis. Input prices and capital outlays were indexed to a 1993 basis. Finally, the feedstock cost is corn stover harvest cost adjusted for transportation as discussed above. The overall production cost for ethanol is estimated at $0.46 per gallon (refer to Table A-2).

FUEL EFFICIENCY

Miller and colleagues (1996) reported that ethanol reduces fuel efficiency in automobiles by 2 percent when used in the standard 10 percent blending proportions with gasoline. Hence, a gallon of ethanol is worth only 80 percent of a gallon of gasoline to an automobile engine. Thus, for $0.46 per gallon of ethanol, it will cost $0.46 (0.80 = $0.58) to produce the ethanol equivalent to a gallon of gasoline.

B

Biographical Sketches of Committee Members

Charles J. Arntzen (*Co-chair*) is the president and chief executive officer of the Boyce Thompson Institute for Plant Research, Inc., in Ithaca, New York. Dr. Arntzen received his Ph.D. from Purdue University in cell physiology. His research interests include cell physiology, biochemistry and development, photosynthesis and chloroplast biogenesis, and plant biotechnology, especially in the use of plants for drug discovery and production of pharmaceuticals. He has edited five books and the *Encyclopedia of Agricultural Science* and has coauthored more than 160 publications. Dr. Arntzen was elected to the National Academy of Sciences in 1983 and to that of India in 1984.

Bruce E. Dale (*Co-chair*) is professor and chair of the Department of Chemical Engineering at Michigan State University, East Lansing. He also holds an appointment as professor of agricultural engineering at Michigan State University. Professor Dale received his Ph.D. from Purdue University. His research interests include integrated utilization of renewable resources, bioremediation, and identification and elimination of rate-limiting steps in biological systems.

Roger N. Beachy is a holder of the Scripps Family Chair and a member of the Department of Cell Biology at the Scripps Research Institute, La Jolla, California, where he is head of the Division of Plant Biology. He is also co-director of an international science training program for researchers from developing countries, the International Laboratory for Tropical Agricultural Biotechnology. Dr. Beach received a B.S. degree in biology from Goshen College, Goshen, Indiana, and a Ph.D. degree in plant pathology from Michigan State University. Dr. Beachy's interests include plant virology and phytopathology, plant gene expression and agricultural biotechnology. He is a member of a numerous scientific societies and has served on previous committees of the National Research Council in areas of agricultural policy and biotechnology. He was elected to the National Academy of Sciences in 1997.

James N. BeMiller is a professor in the Department of Food Sciences and director of the Whistler Center for Carbohydrate Research at Purdue University. He received his Ph.D. degree in biochemistry from Purdue University. Professor BeMiller's research interest is carbohydrate chemistry, in particular determination of chemical structures, modification of chemical and molecular structures, and structure-functional property relationships of carbohydrate polymers as related to practical applications of carbohydrates.

Richard R. Burgess is a professor of oncology at the McArdle Laboratory for Cancer Research at the University of Wisconsin, Madison. He has also been director of the university's Biotechnology Center since it was started in 1984. He obtained his Ph.D. in biochemistry and molecular biology at Harvard University. Dr. Burgess consults with a number of biotechnology companies and plays an active role in educating the public about biotechnology.

Paul Gallagher is associate professor of agricultural economics at Iowa State University. He received a Ph.D. in agricultural economics from the University of Minnesota in 1983. Dr. Gallagher's research interests include price analysis, policy analysis, and agricultural marketing.

Ralph W. F. Hardy is a plant biochemist with research and management activities in both the nonprofit and for-profit private sectors. Dr. Hardy received his Ph.D. degree from the University of Wisconsin. Until September 1995, he was president and chief executive officer of the Boyce Thompson Institute for Plant Research at Cornell University in Ithaca, New York. He currently serves on the Alternative Agricultural Research and Commercialization Corporation Board of the U.S. Department of Agriculture. Dr. Hardy has volunteered for numerous National Research Council study committees. He has been a member of the NRC's Board on Agricultural and Natural Resources, Board on Biology, and Commission on Life Sciences. Dr. Hardy is currently a member of the National Agricultural Biotechnology Council.

Donald L. Johnson is vice president and director of research and development for the Grain Processing Corporation, Muscatine, Iowa. Dr. Johnson received his Sc.D. in chemical engineering from Washington University. His primary interests are in the utilization and processing of renewable resources for food ingredients and industrial chemicals. Recent emphasis is on developing bulk and specialty chemicals from carbohydrate-containing raw materials. Dr. Johnson was elected to the National Academy of Engineering in 1993.

T. Kent Kirk is director of the Institute for Microbial and Biochemical Technology at the U.S. Department of Agriculture's Forest Products Laboratory in Madison, Wisconsin. He is also professor of bacteriology, University of Wisconsin, and adjunct professor of wood and paper science at North Carolina State University. Dr. Kirk received his Ph.D. in biochemistry and plant pathology from North Carolina State University. He was elected to the National Academy of Sciences in 1988.

Ganesh M. Kishore is director of Ceregen Technology, a unit of Monsanto Company. He received his Ph.D. in biochemistry from the Indiana Institute of Science. Dr. Kishore's research interests include glyphosate tolerance in key agronomic target crops and crop quality with specific emphasis on carbohydrate and lipid improvement.

Alexander M. Klibanov is professor of chemistry and a member of the Biotechnology Process Engineering Center at the Massachusetts Institute of Technology. He received his Ph.D. in chemical enzymology from Moscow University in Russia. Professor Klibanov's research interests include enzyme technology, stability and stabilization of proteins, and nonaqueous biochemistry. He was elected to the American Institute for Medical and Biological Engineering in 1992 and to the National Academy of Engineering in 1993

John Pierce is director of discovery research at DuPont Agricultural Enterprise. In this capacity he directs the efforts of a diverse group of life scientists who are using modern technologies to enhance the food, feed, and industrial materials uses of major agronomic crops. Dr. Pierce received his Ph.D. in biochemistry from Michigan State University.

Jacqueline V. Shanks is an associate professor of chemical engineering at Rice University. She received her Ph.D. in chemical engineering from the California Institute of Technology. Dr. Shanks's research interests include bioengineering aspects of pharmaceuticals production from plant tissue culture, in situ nuclear magnetic resonance spectroscopy for characterization of cellular physiology and metabolism, and phytoremediation.

Daniel I. C. Wang is Chevron professor of chemical engineering and director of the Biotechnology Process Center at the Massachusetts Institute of Technology. Professor Wang received his Ph.D. in chemical engineering from the University of Pennsylvania. His research interests include transport phenomena in animal cell bioreactors, biosensors in bioprocess monitoring and control, protein purification and protein refolding in downstream processing, bioreactor design in viscous fermentations, and oxygen transfer in fermentation vessels. He served on the Board on Chemical Sciences and Technology, the Board on Biology's standing Committee on Biotechnology, and the National Academy of Engineering's Bioengineering Peer Review Committee; he was elected to the National Academy of Engineering in 1986.

Janet Westpheling is an associate professor of genetics at the University of Georgia. She received her Ph.D. in genetics from the John Innes Institute in England. The primary focus of Dr. Westpheling's research is control of gene expression in the bacterium *Streptomyces* with emphasis on the study of genes involved in morphogenesis and secondary metabolism, the control of carbon utilization, cellular physiology, and primary metabolism.

J. Gregory Zeikus, is president and chief executive officer of MBI International (formerly the Michigan Biotechnology Institute). He received his Ph.D. from Indiana University. In 1992 Dr. Zeikus received an honorary degree (Doctor of Honoris Causa) from the University of Ghent, Belgium. Dr. Zeikus's research interests focus on the biocatalytic synthesis of organic compounds under extreme environmental conditions of temperature and substrate product concentration.